Finding a Sense of Place

Finding a Sense of Place

An Environmental History of Zena

Edited by Bob H. Reinhardt

POLEBRIDGE PRESS
Salem, Oregon

Copyright © 2013 by Bob H. Reinhardt

All rights reserved. Printed in the United States of America. No part of this book may be used or reproduced in any manner whatsoever without written permission except in the case of brief quotations embodied in critical articles and reviews. For information address Polebridge Press, Willamette University, 900 State Street, Salem, OR 97301.

Cover and interior design by Robaire Ream

Library of Congress Cataloging-in-Publication Data

Finding a sense of place : an environmental history of Zena / edited by Bob H. Reinhardt.
    pages cm
 Includes bibliographical references.
 ISBN 978-1-59815-128-2 (alk. paper)
 1. Human ecology--Oregon--Zena Forest--History. 2. Zena Forest (Or.)--Environmental conditions. 3. School farms--Oregon--Zena Forest. 4. Willamette University--History. 5. Place-based education--Oregon--Zena Forest. I. Reinhardt, Bob H., 1978- editor of compilation.
 GF504.O7F56 2013
 304.209795'3--dc23

                                                    2013009163

Contents

Illustrations vii

Acknowledgments viii

Contributors ix

Abbreviations x

Introduction: Finding Our Way through Zena
Bob H. Reinhardt 1

1. Zena's Genesis
   Aaron Jackson with Lettajoe Gallup 11

2. Stories of Place
   Nickolas Lormand with Larissa DeHaas 21

3. Kalapuyan Interactions with the Land
   Brayton Noll with Summer Tucker 31

4. Kalapuyan Society and Sense of Place at Zena
   Elena Crecelius 41

5. The Legal Development of Zena
   Michael Harder with Alec Weeks 49

6. Landscape and Religion in Nineteenth-Century Settlement
   Vera Warren with Emily Dougan 57

7. From Wilderness to an Agricultural Landscape
   Amanda McClelland with Andrew Spittler 65

8. Transformative Agricultural Technologies at Zena
   Emily Schlieman with Keller Cyra 75

9. Evolution of Land Use Planning in Oregon
   Kyle Carboni with Morgan Gratz-Weiser 85

10. Sarah Deumling at Zena Forest
    Lauren Henken 95

11. A Bureaucratic Sense of Place
   Philip Colburn with Kevin Bernstein **107**

12. Willamette University at Zena
   Erica Jensen with Erik Sandersen **117**

13. Personal Stories of Zena
   Elise McGlone and Lauren Vermilion **127**

14. Stories of Zena from the Willamette Community
   Heather Smith with Elise McGlone **141**

Epilogue
   Bob H. Reinhardt **151**

Works Cited **155**

Illustrations

1. Map of Zena area. *Courtesy of Trout Mountain Forestry*
2. Spring 2012 Environmental History of Zena students, with Anne Walton and her dog Griffin. *Courtesy of Elise McGlone*
3. Fall 2012 Environmental History of Zena students, with Karen Arabas. *Courtesy of Bob Reinhardt*
4. A view of the Eola Hills from Zena. *Courtesy of Lettajoe Gallup*
5. Map of Kalapuya territories, as reconstructed by Melville Jacobs. *Melville Jacobs,* Kalapuya Texts *(1945)*
6. Acorns found at Zena. *Courtesy of Larissa DeHaas*
7. Yarrow found at Zena. *Courtesy of Larissa DeHaas*
8. 1852 surveyor's map showing Zena-area township. Sanford Watson's claim is highlighted. *Courtesy of the Bureau of Land Management*
9. Map of tile drainage at Zena; the diagonal lines on the left side of the photo represent the location of the tiles. *Courtesy of National Resource Conservation Service, Dallas, Oregon*
10. Farming in the Zena area, 1911. *Courtesy of Anne Gilbert Walton, current resident of Zena Springs Farm, Zena, Oregon*
11. Polk county land use, past and planned. *Polk County Comprehensive Plan (June, 1974)*
12. Sarah Deumling and her dog, Henry. *Courtesy of Lauren Henken*
13. Summer Institute in Sustainable Agriculture students trellising tomatoes in greenhouse, 2012. *Courtesy of Jennifer Johns*
14. Willamette student Kelsey Copes-Gerbitz taking a tree core sample. *Courtesy of Willamette University*
15. Daffodils at Anne Walton's property. *Courtesy of Elise McGlone*

Acknowledgments

The contributors and editor would like to thank all those who gave generously of their time, insight, and stories: Karen Arabas, Joe Bowersox, Kelsey Copes-Gerbitz, David Craig, Sarah Deumling, Rebecca Dobkins, Andries Fourie, Bob Feldman, Kristen Grainger, Gunnar Gundersen, Melissa Hage, Peter Henry, David Lewis, Briana Lindh, Marc Marelich, Pedro Martinez, Eleanor Miller, Matthew Nelson, Bill Olson, Michael Pope, Scott Pike, Kari Ramey, Sue Reams, TJ Sandvig, Zack Taylor, Laura Tesler, Steve Thorsett, Anne Walton, and Karl Weist.

Special thanks to Willamette University's Center for Sustainable Communities, which provided funds for the publication of this book through a Faculty Fellow Research Grant.

Contributors

Spring 2012 Class
  Kyle Carboni (Environmental Science)
  Philip Colburn (Environmental Science)
  Elena Crecelius (Environmental Science)
  Keller Cyra (Environmental Science)
  Michael Harder (History)
  Lauren Henken (Environmental Science)
  Aaron Jackson (Environmental Science)
  Erica Jensen (Environmental Science)
  Nickolas Lormand (Undeclared)
  Amanda McClelland (Environmental Science)
  Elise McGlone (Undeclared)
  Brayton Noll (Environmental Science)
  Vera Warren (Psychology)

Fall 2012 Class
  Kevin Bernstein (Environmental Science and Politics)
  Larissa DeHaas (Environmental Science)
  Emily Dougan (Environmental Science)
  Lettajoe Gallup (Environmental Science and Politics)
  Morgan Gratz-Weiser (Environmental Science)
  Erik Sandersen (Undeclared)
  Emily Schlieman (History)
  Heather Smith (Environmental Science)
  Andrew Spittler (Environmental Science)
  Summer Tucker (Environmental Science)
  Lauren Vermilion (Politics)
  Alec Weeks (Environmental Science)

Abbreviations

ASP	American Studies Program
BPA	Bonneville Power Administration
DDT	dichloro-diphenyl-trichloroethane
DLCD	Department of Land Conservation and Development
ESA	Endangered Species Act
EPA	Environmental Protection Agency
FSC	Forest Stewardship Council
LCDC	Land Conservation and Development Committee
NEPA	National Environmental Policy Act
NWPA	Northwest Power Act
NWCC	Northwest Power Conservation Council
ODFW	Oregon Department of Fish and Wildlife
DLC	Oregon Donation Land Act
PCCP	Polk County Comprehensive Plan
SCS	Scientific Certification Service
TIUA	Tokyo International University of America
TMF	Trout Mountain Forestry
TPL	Trust for Public Land
USDANRCS	United States Department of Agriculture Natural Resources Conservation Service
WU	Willamette University
ZSI	Zena Sustainability Institute

Introduction

Finding Our Way through Zena

Bob H. Reinhardt

You would be forgiven a few moments of bewilderment on your first visit to Zena Forest and Farm. Turning off Zena Road and onto a short gravel driveway, the first thing you see is an old farmhouse, which is not atypical for this part of the Willamette Valley. The small barn to the side of the house isn't all that odd, either, although the much larger greenhouse next to the barn might strike you as a bit strange. But when you really start looking around, Zena quickly becomes a very confusing place. Depending on the time of year, you might see some people deliberately starting fires; if you asked, you would learn they're burning not to prepare for an agricultural crop, but to restore an environment—oak savanna—that dominated the area centuries ago. There is some farming here, and although it's organic and relatively small, some time-worn and scraggly fruit trees suggest more commercial ambitions long ago. And what's that odd white structure on the hill—a telescope? Is that the sound of chainsaws in the distance? And then there are the college students: some in hard hats and rubber boots, others with tree-coring tools and notebooks, and still others sitting in a circle for a discussion while another group swirls and bounces in a choreographed dance. Who are these people? What are they doing? What does all of this mean?

This book represents a collaborative effort to explore that confusion and begin to make sense of Zena. The outlines of that place are easy enough to define. The 305-acre property sits in the Eola Hills of the Willamette Valley, ten miles northwest of the Willamette University campus in Salem, Oregon. Willamette purchased the parcel in 2008, with the intent of both developing educational programs and "protecting, restoring, managing, and enhancing the natural resources and ecosystem services of Zena." Some detailed and fascinating studies by Willamette faculty and students—

2 Introduction

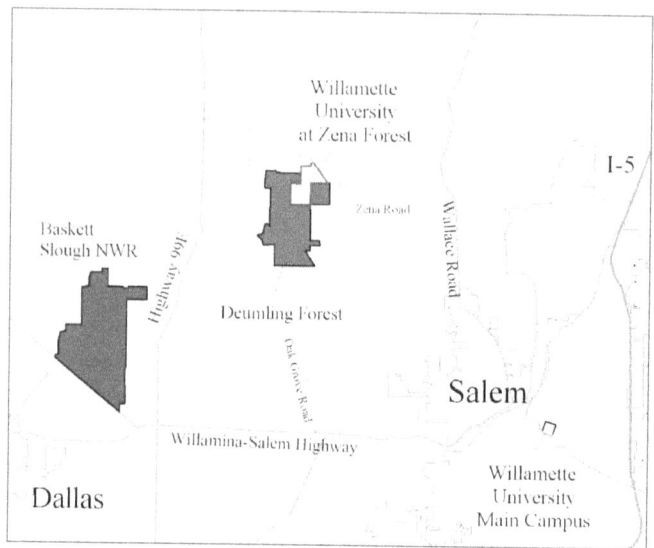

1. Map of Zena area. *Courtesy of Trout Mountain Forestry*

especially Kelsey Copes-Gerbitz—have outlined the property's pre-Willamette history. Zena lies within the traditional land of the Luckiamute band of the Kalapuyan peoples, now part of the Confederated Tribes of Grand Ronde. The United States government took possession of most native land in the Willamette Valley through the Kalapuya Treaty of 1855, and then distributed what is now the Zena area to three Euro-American settlers (Sanford Watson, Thomas Warriner, and Edward Robson). They, their descendants, and later buyers of the property used the area for both agricultural and monocultural (that is, single-species, Douglas fir) forestry purposes until 1984, when the German count and investor Hermann Hatzfeldt began buying up Zena for more diverse forestry practices. Hatzfeldt's total holdings—what is properly called "Zena Forest"—eventually amounted to more than 2,000 acres including the 305 acres that Willamette purchased in 2008.[1]

In short, we know quite a bit about what Zena is and how it has changed—but we know much less about why. That was the question taken up by a rather unusual course called The Environmental History of Zena, offered at Willamette University in the spring and fall of 2012. During those two semesters, twenty-five students (thirteen in the spring, twelve in the fall) worked together to conceptualize, research, and write the first coherent and comprehensive—though not to say complete—history of Zena. The result is a remarkable book that not only stretches from Zena's geological past to plans for the property's future, but also explores a variety of methods, tools, and themes as a way of understanding a truly remarkable place.

The book and the course from which it came start from the proposition that environmental history helps us tell better—that is, more accurate, detailed, and meaningful—stories about the past. Environmental history, in general, focuses on the relationships between humans and "nature"—more specifically, the nonhuman natural world—as they have changed (and not) over time. Such stories often deal with how humans have changed their environments—cutting down trees, damming rivers, etc.—but they also explore how the nonhuman natural world accommodates, resists, and alters such efforts, sometimes as a real physical force (climatic forces, for instance) and sometimes as an idea (such as the aesthetic value of natural scenery). This approach to studying and understanding the past requires a fundamentally interdisciplinary approach, incorporating the tools and methods of the sciences and the humanities. In practice, this environmental history project required students to draw on a variety of sources: studies of the Willamette Valley's geology, oral histories from the Kalapuyan people, church records and state archival materials, interviews, and more. Each of these sources provided some insight into changes and continuity in interactions between humans and the nonhuman natural world at Zena.[2]

In an effort to bring together those disparate insights and stories about the past, this book uses and contributes to a fascinating analytical framework: "sense of place." The anthropologist and ethnographer Keith Basso has provided the most sustained treatment of this concept, particularly through his study of the Western Apache, *Wisdom Sits in Places*, and as co-editor of the collection *Sense of Place*. Other scholars from different disciplines have also seized on the idea: geographer Yi-Fu Tuan, historian William Lang, and Western writer Wallace Stegner, to name just a few. These and other scholars have been drawn to sense of place because of the concept's interpretive power to explain how people connect to, understand, and interact with their environments. And that is precisely why this book uses sense of place as its unifying theme: a broad concept that allows exploration of the myriad reasons and ways in which people have changed, and been changed by, Zena.[3]

The very breadth that gives sense of place its conceptual power also produces analytical challenges; the concept comes without a clear set of rules and vocabulary (or dogma and jargon, depending on one's perspective). And so the authors of this book not only use sense of place to explore Zena's past, but also use Zena to explore sense of place—to develop their own ideas, understandings, and applications of the concept. Taken together, these disparate analyses offer two important contributions to this analytical framework. First, the book explores the variety of ways in which a person or group's sense of place begins and evolves through their interactions with the land: the *origins* of a sense of place. Second, the authors show that sense of place matters in shaping how people change the land: the *effects* of sense of place. In so doing, the book demonstrates the utility of the concept of sense of place for understanding why and how people interact with their environments (the origins and effects, respectively), in the past, present, and future.

The decision to use sense of place came from the student-authors themselves, who set the path for this exercise in student-led place-based pedagogy. Except for a few weeks of introductory material about the practice of environmental history and an outline of the Willamette Valley and Zena's past, the course and project evolved under student leadership. The students defined the scope and content of the book: when it should start, when it should end, and what topics it should cover, making difficult choices about what could and could not be accomplished. For instance, the students decided to expand their investigations beyond Willamette's 305-acre parcel to the larger Zena Forest, Eola Hills, and even the Willamette Valley more generally, partially in response to the paucity of Zena-specific sources, but, more importantly, in order to draw the connections necessary to understand the history of Willamette's property. The students then decided who would be responsible for each chapter—also a challenge, given passionately-felt and common interests—and established expectations for the book, from the depth of research and historiography to the quality of prose. The students planned their research agendas, outlined their chapters, and reviewed and edited each others' work. Because two different classes worked on the project over two semesters, some unequal divisions of labor were unavoidable: for instance, the first semester's students defined the scope of the book and wrote the first draft, while the second semester's students produced the final draft with-

2. Spring 2012 Environmental History of Zena students, with Anne Walton and her dog Griffin. *Courtesy of Elise McGlone*

out much input from the first class. But the focus of the course on Zena keeps the project from spinning away in too many disparate directions. In this, the course (and book) was also an exercise in place-based pedagogy, in which academic inquiry centers on a particular place—in this case, Zena Forest and Farm and the surrounding area. Environmental historians have implemented this approach to particularly good effect as a teaching tool, producing fascinating histories of the Wicomico River in Maryland and the Peace River in British Columbia. Place-based pedagogy provides both students and teachers the opportunity to extend their historical investigations out to national, regional, and global themes and events, all while maintaining their focus on a particular point of reference. With a common object of study and shared responsibility for the outcome, the twenty-five student-authors produced a text that, through many different voices, moves towards one goal: telling the best possible story about Zena's past.[4]

This book tells that story both chronologically and thematically. It begins with an introduction to Zena's geological history, written by Aaron Jackson and Lettajoe Gallup. These students faced a unique challenge: how to do environmental history—which focuses on the nonhuman *and* human—without any humans. The chapter surmounts this problem by both eschewing environmental determinism, in which people have no choice but to do what the nonhuman natural world allows them, and showing how humans and their sense of place would become woven into an already interconnected and evolving world.

The next three chapters focus on the Kalapuyan people, the original human inhabitants of the Willamette Valley. Although no conclusive archaeological evidence has yet been found that incontrovertibly proves the presence of the Kalapuya on the 305-acre Zena Forest and Farm, the area lies within traditional Kalapuyan grounds, and, as Brayton Noll and Summer Tucker's chapter reveals, contains (or once contained) many plants and animals used by the Kalapuya. Tucker and Noll argue that the concept of Traditional Ecological Knowledge (TEK) provides a framework for understanding how and why the Kalapuya used what might now be called "natural resources," and how such knowledge stemmed from and informed their connection to the nonhuman world. Nickolas Lormand and Larissa DeHaas also investigate interactions between the Kalapuya and their environments, using a selection of Kalapuyan stories about origins, natural phenomena, and more. DeHaas and Lormand suggest that such stories provide not only a way of making sense of the nonhuman natural world and the Kalapuyan place in it, but also a way for people to connect to each other through the environment—a provocative idea for our own time, as the chapter concludes. The section finishes with Elena Crecelius's chapter on Kalapuyan society and some of its structures and practices. Crecelius observes the variety of ways in which Kalapuyan society was deeply and intimately connected to the nonhuman natural world—so deeply, Crecelius argues, that the destabilization of ecological conditions (caused by the presence and actions of Euro-Americans) had a similarly destabilizing effect on nineteenth-century Kalapuyan society.

The second section of the book continues the story of Euro-American migration and settlement in the Zena area and Willamette Valley more broadly, moving from the mid-nineteenth century to the mid-twentieth century. Michael Harder and Alec Weeks begin the section with an account of the various legal mechanisms used by Euro-Americans to officially possess and bound the Zena area. More importantly, Weeks and Harder argue that these mechanisms made it possible for Euro-Americans to conceptualize and understand Zena, and to develop a connection to that place. Of course, westward-bound migrants and settlers drew from other sources in their conceptualization of Zena; as Vera Warren and Emily Dougan show, a particularly Protestant vision of Eden informed settlers' ideas about what the landscape of Zena was, could, and should be. But Dougan and Warren also reveal how that landscape shaped those religious visions and ideas, so that, for instance, the songs and prayers of the community's church bore the imprint of the natural forces buffeting Zena-area congregants. Amanda McClelland and Andrew Spittler's chapter continues the focus on the nineteenth-century Euro-American community, focusing on three individuals in particular: Sanford Watson, Thomas Warriner, and Edward Robson, the first Euro Americans to hold legal title to the land that would eventually become Willamette's 305-acre property. Combining the rare fragments of archival material specific to these individuals with more general information about Willamette Valley settlement, Spittler and McClelland highlight ways in which the nonhuman natural world challenged Zena-area farmers, and how those farmers confronted such challenges. Their chapter also serves as a helpful reminder that nineteenth-century farmers were, in fact, quite oriented towards market exchange, a production- and profit-mindset that continued into the twentieth century, as Emily Schlieman and Keller Cyra show in their chapter. Schlieman and Cyra use the fascinating example of tile drainage—subsurface pipes installed to make wet soil sufficiently dry for particular kinds of agriculture—to reveal how science and technology, particularly after World War II, gave farmers ever more powerful tools to manipulate, but never quite master, the nonhuman natural world.

Such efforts at mastery would produce reaction and backlash, as the next section of the book shows. Kyle Carboni and Morgan Gratz-Weiser outline the history of land use planning in Oregon, with a particular focus on Senate Bill 100, Oregon's first-in-the-nation land use planning law, and its effects on Polk County (in which Zena sits). Gratz-Weiser and Carboni also detail some of the conflicts that land use planning prompted, particularly between urbanites—concerned about water quality, recreation opportunities, and natural aesthetics—and rural residents, including those in the Zena area, whose sense of place is shaped by their lives on the land and the livelihood they derive from it. Among those living on—in so many different ways—the land at Zena is Sarah Deumling, the subject of Lauren Henken's chapter. Henken explains how Deumling came

3. Fall 2012 Environmental History of Zena students, with Karen Arabas.
Courtesy of Bob Reinhardt

to own and manage Zena Forest, from which Willamette acquired its 305-acre parcel. More than that, Henken provides a stimulating and moving account of how Deumling connects to the land, and how that sense of place shapes her actions as a landowner.

While Henken's chapter portrays an intimate sense of place, Philip Colburn and Kevin Bernstein's chapter turns towards disembodied and distant bureaucracies—specifically, the Bonneville Power Administration and the Trust for Public Land—that made the Deumling-Willamette land transfer possible. Bernstein and Colburn explain how a relatively obscure mechanism—a conservation easement—provided the necessary funds for Willamette's purchase of Zena, a story that the chapter situates in the broader history of post-World War II American conservation, preservationism, and environmentalism.

The book concludes with three chapters about different views of, objectives for, and stories about Willamette University at Zena Forest. The section begins with a detailed look at Willamette University's decision in 2008 to purchase Zena and go into the business of sustainable forestry, restoration ecology, place-based pedagogy, and . . . well, who knows what else? Erica Jensen and Erik Sandersen's chapter relates these and other ideas about what Zena can and should be used for, suggesting the potential for tension between those visions—unless the Willamette community recognizes the different senses of place that spur such passionate ideas. Elise McGlone and Lauren Vermilion bring our attention to other passionate ideas about Zena by taking us away from the Willamette University community. Their chapter allows five "outsiders" to make their own suggestions about what Willamette ought to do with its 305-acres—ideas which, as Vermilion and McGlone show, reflect those observers' own sense of place. The final chapter brings the book back inside the Willamette University community and reveals a range of visions for Zena's future. Heather Smith and Elise McGlone explore Zena's role in fulfilling the Willamette University mission by recounting a variety of different stories about what Zena is and can be. Each of these stories comes from a particular sense of place, and each depicts a different role for Zena in recognizing that, as the Willamette motto enjoins, "not unto ourselves alone are we born."

That the book ends with a variety of stories about Zena is appropriate, for the book itself is a story about Zena. And it should be treated as such: as *a* story—not the story—of Zena's past. This book is not meant as the definitive, authoritative, or final account of Zena's history. Just the opposite, in fact; this book came from an evolving collaboration, and such an evolution of Zena's story and meaning should continue. That ongoing process of discovery will undoubtedly reveal areas in which this book is lacking. But those shortcomings are easily outweighed by the many contributions of this book: its range of topics, its variety of research material (both familiar and original), and its myriad of tools, methodologies, and analytical approaches. Ultimately, in its stories about Zena's past, this book helps us begin to understand the meaning of Zena, and to develop our own sense of that remarkable place.

Notes

1. "The University Forest at Zena: Mission and Goals"; Copes-Gerbitz, "Defining the Historical Context of Zena Forest, Salem, Oregon." Willamette University's Center for Sustainable Communities offers more information about Zena at its website: http://www.willamette.edu/centers/csc/zena/. Willamette's Earth and Environmental Sciences department has a website listing recent senior theses, including some of those about Zena: http://www.willamette.edu/cla/ees/senior_papers/index.html.

2. Environmental history has developed a vast and sprawling literature since the 1960s and 1970s; for a recent survey of the field and its uses in the classroom, see the October 2011 issue of *The Organization of American Historians Magazine of History*.

3. Basso, *Wisdom Sits in Places*; Feld and Basso, *Senses of Place*; Lang, "From Where We Are Standing"; Stegner, *Where the Bluebird Sings to the Lemonade Springs*; Tuan, *Space and Place*.

4. Lewis, "'This Class Will Write a Book'"; Evenden, "Reflections: Environmental History Pedagogy."

1

Zena's Genesis

Aaron Jackson with Lettajoe Gallup

The sheer overcast of light grey clouds cover the sky, while drizzle splatters the leaves below. The landscape presents the sight of rolling hills sectioned off into farm plots, a forest filled with tall Douglas firs, and developing oak savannas. As the rain starts to pour, a small but powerful stream cuts through thick vegetation that collects the water running down the hillside. Mud, mixed with a variety of sediments, small rocks, and fallen vegetation, sticks to the sides of shoes. Tall tan grasses stand in the way; blackberry bushes with plump fruit tangle with other plants on the side of the trail. Further along the trail there is open field, cleared by burning practices. Mounds of smoking ash are scattered over the area on a bed of standing brown grasses. Back at the farmhouse, a young kitten waits, sitting erect on the porch and protected by the rain. Taking one last look at this forest, *serenity* is the only word that comes into the mind.

Beneath all of this sublime beauty is a long and complex history of hills, streams, soils, plants, and animals. The story begins underwater, 80 to 50 million years ago, when North and South America took their modern positions relative to the other continents. At the time, a tropical sea extended as far as the Cascades, submerging Zena along with the rest of the Pacific Northwest. Underneath that sea, basalt volcanic islands grew in a scattered distribution. Around 30 million years ago, tectonic activity rumbled at the border between continental and oceanic plates, causing the sea to recede. The folding and faulting movements of plates spurred volcanic activity, lifting and forming the modern Coast Range and producing the heavy basalt flows that are evidence of the volcanism. Calcium-rich sediments, mostly basalt in nature, then filled in what is the modern-day Willamette Valley, which was still submerged underneath the receding ocean. The Coast Range Mountains would grow gradually during the next 25 million years to reach their current extent and height. The sea withdrew west from the land 23.7 million years ago, exposing the volcanic islands; approximately 22 million years later, the sea gradually retreated further west to its modern day location. The disappearance of the inland sea gave way to the modern stream patterns, such as the Spring Valley and Richards creeks that spread throughout Zena Forest, within the forming valley. Ten million years ago, a significant basalt flow obstructed the northern outlet of the valley from the flow of any stream tables. Because the paths of the rivers formed 23.03 to 5.332 million years ago were blocked by geologic activity, the Willamette Valley experienced a great deal of backflow. The basalt flow contributed to form a fifty-foot cliff into the Willamette Valley, affecting where the rich alluvium spread at the bottom of the waterfall.[1]

In short, it took millions of years and a variety of intertwined forces and processes to create Zena's hills, streams, soils, and vegetation. Each force and process spurred another one, like a huge chain reaction. The impressions and emotions inspired by the Zena landscape, and the ways in which people connect to and develop a sense of Zena—those, too, are part of the chain reaction of interconnected processes stretching back millions of years.

Hills

From the platform where a medium-sized telescope sits, a visitor can gaze out towards the surrounding Eola Hills and beyond. Most of the hills gradually slope on the eastern side and have a steep decline on the western side. The highest hills range from 850 to 1163 feet. Some hills have farmland on one side, forest on the other, and green grassy hillsides connecting them both. To the north, the Amity Hills present themselves, large in size but few in number. Eastward lie acres of luscious farmland in the East Valley Plain. To the south, the Willamette River winds through farms and fields, and to the west, small mounds and hills are sprinkled throughout the valley plain.[2]

The Eola Hills expand 50 square miles from west Salem, nearly reaching the town of Amity, five miles to the north of Zena. The hills are a product of volcanic activity in the western Cascade region that reached a peak between 16.6 and 11.2 million years ago, while the sea was still retreating west. That activity created the Columbia River basalt flows, which covered more than 63,000 square miles, from an eastern border stretching between today's Spokane to Boise, then narrowing on the way west to the Pacific

4. A view of the Eola Hills from Zena. *Courtesy of Lettajoe Gallup*

Ocean, while dripping south into the Willamette Valley and today's Zena. The hot lava cooled and crystallized, and was then whittled down by the wet climate. The water eroded the volcanic landscape, leaving dark red soils to surround the lava caps. Isolated lava caps are scattered between Salem and Portland; these lava caps are the foundation for the Eola Hills.[3]

Eola stands for the goddess of the winds—another of the interconnected forces that created the hills of the Zena area. Over time, climate forces wore down the high peaks of lava caps, transforming them into rolling hills. Those hills, like the rest of Zena, are a product of multiple systems working in conjunction with one another. The processes of faults and earthquakes caused the majority of the volcanic activity that created lava caps. The weather solidified and eroded the hills, leading to a particular elevation profile that, in turn, shaped the particular configuration of vegetation, soils, and climate at modern day Zena. All of these forces are still at work—slow and unseen, but moving in and under the hills of Zena, even as visitors form their impressions, emotions, and sense of an imperceptibly-shifting place.

Hydrology

Little wet droplets fall from the sky. They land on the leaves of the trees, the soil on the ground, and the different platforms formed by the various vegetation. Rain splashes the surface of the Spring Valley Creek, creating a loud rush of water through the stream's path. These sounds can be heard from the trail, but the stream is only accessible through the brush. While walking downhill 200 feet to the stream, one feels the soil closer to the stream softening underfoot as groundwater collects underneath in large pockets of space. The stream contains cold water with a foggy tint, littered with dirt, dead leaves, and branches. The narrow stream looks like it travels for miles, cutting through vegetation and winding around trees before eventually depositing into another body of water well beyond view.

Two above-ground stream systems originate at Zena: Spring Valley Creek and Richards Creek. Spring Valley Creek has three branches; Richards Creek has two. These streams and branches all take shape in a similar fashion. Water from rain or snow/icemelt that does not seep into the ground, flows downhill over the surface. Gradually the surface water creates channels by eroding the top layer of sediments. These channels combine to form a larger stream. Since these stream systems are located on a hill, all the channels eventually meet at some point towards the bottom.[4]

The stream systems help create groundwater. The water that does not get deposited by the streams finds its way through the permeable soils, then through cracks and holes in the rocks. The water slowly decomposes the surface rocks by wearing down the molecular structure, creating larger passageways for the water to continue traveling deeper into the earth. Eventually the water pools in a pocket filled with sandstones, trapped above an impermeable rock layer. The water moves through the fine-grained sandstone, creating a pocket of sandstone and water called an aquifer. The groundwater level at Zena is between 120 and 140 feet in elevation, and the highest elevation of Zena is 662 feet — which is to say that the water traveled through hundreds of feet to reach the water table.[5]

Water constantly enters the forest through rain and the different streams that feed into Zena, and it regularly leaves the area through evaporation and transpiration. Because weather patterns change daily, Zena's hydrology is constantly changing. With the extent of rain Zena experiences, the channels are constantly expanding. Spring Valley Creek floods during heavy periods of rain. Over time this flooding expands the creek by eroding the walls of the waterway. This transformation has been happening for thousands of years, and it will continue to happen, constantly shifting the hydrology of the landscape.

Water, then, moves above, on, and below Zena, and it has always been an essential component of the landscape. Zena used to be the floor of a sea; then the presence of water disappeared for a couple million years during the vicious volcanic activity. Eventually the water came back to help cool and crystalize the lava; later, water came in full force during the Missoula Floods. And now Spring Valley Creek and Richards Creek cut through Zena, burbling and churning on the land, in the ear, and through the mind.

Sediments

The ground at Zena is composed of an array of sediments that have accumulated over thousands of years. The top layer consists of dead leaves that have fallen from the trees. Under the leaves are numerous thick grasses dead and living. There are some bare spots that just have a coloring of light brown with grey mixed in. The dirt holds on to the bottom of shoes; imprints are left by people, vehicles, and animals. Underneath the top layer of the soil lie many more layers, composed of different materials with origins dating back millions of years.

The sediments have gone through many different processes to be as fertile as they are today. Two million years ago, large ice sheets covered extensive portions of northern Asia, Europe, and North America, including Zena, buried at that time under an ice sheet that filled valleys throughout northern Washington, Idaho, and western Montana. One million years ago, alpine glaciers formed on the eastern mountains, and during interglacial periods and the short summers of the glacial period, the melt water from these alpine glaciers contributed to valley streams feeding into the Willamette and the Columbia rivers. The streams from the alpine glaciers' melt carried alluvial material to deposit it into the paths they cut into the valley. The retreat of the ice sheets and the erosion of the alpine glaciers during the end of a glacial period resulted in the inundation of the Willamette Valley with lakes that reached depths up to 100 feet, though these lakes had greatly evaporated by the advance of the next glacial period 60,000 years ago.[6]

One particular continental glacier created a significant force that would impact the sediment disposition of the Willamette Valley. Glacial Lake Missoula formed in western Montana around 15,000 years ago; it covered 7,700 square kilometers and contained about 2,100 cubic kilometers of water, locked behind glacial ice walls that dammed off the water from lowland environments. At the end of the glacial period, longer summers undermined the ability of the glacial ice to sustain the great lake, and the inevitable breach of the dam led to the reintroduction of the lake's water content back into the environment. Between 15,000 and 13,000 years ago, Lake Missoula's glacial dams broke up, causing multiple floods throughout the surrounding basins. Much of the Willamette Valley's modern topography, including the variations in elevation along Zena's trails, owes its formation to the flow of the flood across the landscape. Some large and foreign-in-composition boulders called erratics—dramatic evidence of the power the Missoula Floods—are scattered throughout the Willamette Valley and the scablands of the Columbia Basin. The closest erratic to Zena is at Erratic Rock State Natural Site, thirty-three miles northwest.[7]

The Willamette Valley also owes part of its fertility to this event, which formed soil favorable for the growth of native species. But that fertility is the result of other processes, too, including the basalt flows. Sediments exemplify how the environment changes as time goes on. Each sedimentary layer developed at different times, one after another, piling up to create meters of rich soil. That soil continues to change as vegetation decomposes, mixes, and is compacted into existing sediments. The ground underfoot at Zena, then, is the product of forces both historical and contemporary, as are the ways in which visitors see and understand Zena.

Paleoenvironments

On an average late autumn day, visitors to Zena are greeted with an overcast sky and chilly temperatures, sometimes with a little bit of sun trying to poke through a constant light drizzle. The day begins with the sun rising over a heavy layer of dew. A fine rain takes over in the midmorning then turns into pouring droplets in the afternoon. The wind guides the rain in a diagonal direction while it falls to the ground. As the day goes on, the rain settles into an alternating mist and sprinkle of rain. The bark of the trees is soaked through with wetness. In the distance, fog rises from the hillsides then disappears into the atmosphere.

From October through May the average precipitation in the Willamette Valley is 110 centimeters. Throughout the rest of the year there is infrequent precipitation as well as a mild drought in the summer. During the summer, particularly in July, the average temperature is nineteen degrees Celsius, whereas in January the average is 4.5 degrees Celsius. But the scene was much different 50 million years ago. Then, the continental mass presented a warm tropical setting, not the marine coastal environment as it is today. From 36.6 to 23.7 million years ago the climate started to shift from a tropical to a warm temperate environment. Towards the end of the basalt floods and volcanic activity from 5.3 to 2 million years ago, the climate cooled and changed to a more semi-arid condition. The temperatures rapidly decreased as ice sheets formed in the north from 2 million to 1500 years ago, during the Ice Age. Temperatures then increased, creating catastrophic floods throughout the valley. During the "Little Ice Age" (1450–1850 CE), the climate shifted again, towards more precipitation and cooler temperatures.[8]

Climate plays a crucial role in a host of processes, from what kinds of plants can survive to what kinds of lives people can lead. Today, climate shapes a variety of activities at Zena: sustainable agriculture, prescribed burning, even field trips. Humans respond to shifts in the climate, just as the landscape itself was shaped by paleoenvironments—another example of the deeply interconnected nature of Zena.

Vegetation

Zena's color palette consists of a variety of greens and different shades of brown. The different species of trees grow amongst each other: skinny and short oaks with sparse limbs; tall and thick Douglas firs with long branches. Deeper into the woods, trees lean against the slopes with lichens hanging off the limbs and fungi growing at the base. On the side of the hill great sword ferns stand out from all the grasses, growing in packs with one very close to the other, their undersides embellished with brown seeds to keep their line growing. Towards the stream, some trees look sad—almost sickly—from drowning in the abundance of water. On higher ground, the trees are more grandiose and enriched with various colored leaves.

The vegetation at Zena comes from ancient roots. About 50 million years ago palms, figs, and avocados flourished in the tropical climate. As the climate changed to a more temperate environment, 36.6 to 23.7 million years ago, different species, including magnolia and dawn redwood (metasequoia) trees, appeared. When the climate cooled, warm-temperature plants gave way to species like Douglas fir (*Pseudotsuga menziesii menziessi*), Western hemlock (*Tsuga heterophylla*), and Sitka spruce (*Picea sitchensis*). Remnants of this time period are expressed in low-elevation portions of Zena dominated by mostly Douglas firs and other plant life that adapted to moister climates. The well-drained foothills bordering the Willamette Valley, however, saw the dominance of Grand fir (*Abies grandis*) and Ponderosa pine (*Pinus ponderosa*).[9]

This array of tree and other plant species, like all of the Zena landscape, is the product of a convergence of particular forces and processes. First there had to be the appropriate setting, such as elevation of moderate height. Fertile sediments had to be spread throughout the land. The plants required other nurturing forces, like winds to transport the seeds of the trees, ferns, and grasses, and particular temperatures and humidity levels to encourage the growth of different kinds of vegetation. Zena's plant life illustrates the extent of interconnected systems, as well as how each stage in the development of this land created a different place for the appropriate species to flourish. Ironically, this complex and in many ways contingent array of species has created elements of a landscape that for many visitors seems permanent and essential to what Zena was, is, and might yet be.

Conclusion

Zena arouses a variety of feelings for different people. Some see it as sacred land; others see it as a profitable enterprise. But underneath those impressions is a rich history of interconnected processes. Those processes transformed Zena from molten lava to luscious hills, creating a particular environment upon which people would live, work, and sense in their own ways.

Notes

1. Allen, *The Magnificent Gateway*, 34, 44; Alt and Hyndman, *Northwest Exposures*, 148–66; Glenn, "Late Quaternary Sedimentation and Geologic History," 153.

2. Price, *Ground Water in the Eola-Amity Hills Area*, 10; Price and Johnson, *Selected Ground Water Data*, 3.

3. Allen, *The Magnificent Gateway*, 19, 37–38; Orr et al., *Geology of Oregon*, 206; Price, *Ground Water in the Eola-Amity Hills Area*, 20.

4. Leopold, *Water, Rivers and Creeks*, 39–40; Sims, *Forest Management Plan*, 16.

5. Leopold, *Water, Rivers and Creeks*, 17–18; Price and Johnson, *Selected Ground Water Data*, map.

6. Alt and Hyndman, *Northwest Exposures*, 361.

7. Aikens et al., *Oregon Archaeology*, 284; Denlinger and O'Connell, "Simulations of Cataclysmic Outburst Floods."

8. Allen, *The Magnificent Gateway*, 34, 44; Alt and Hyndman, *Northwest Exposures*, 148–66; Walsh et al., "1200 Years of Fire and Vegetation History," 275.

9. Allen, *The Magnificent Gateway*, 34, 44.

2

Stories of Place

Nickolas Lormand with Larissa DeHaas

Zena is a place where oak, Douglas fir, and ponderosa pine live. It is also a place with a farm house and a small organic farm, tended with care by Willamette students. It is, in other words, a place of both nonhuman and human history. The original human inhabitants, the Kalapuya people, wrote these histories in part through stories that wove together the human and the nonhuman natural worlds. Those stories suggest how the landscape shaped Kalapuyan culture, how Kalapuya actions in turn shaped the land, and, ultimately, how the Kalapuya interacted with and connected to—that is, formed their sense of place with—the Zena area.

In his book *Wisdom Sits in Places*, Keith Basso proposes that people develop their sense of place in part through "communicat[ing] about the landscape"—in other words, by telling stories about place. These stories provide not only the environmental history of a landscape, but also reveal that landscape's interconnection to the culture's system of values. As the historian William Lang explains, these stories about place help create a unique "home . . . a symbolic understanding of place" that weaves identity and history: "the product of narratives about experience and location that often emphasize a distinctive history." Oral narratives help reveal how the Kalapuya "communicated about the landscape" and suggest their understandings of the landscape and their place in it.[1]

This chapter approaches Zena's history through different stories of the Kalapuya as transcribed by Melville Jacobs in the mid-twentieth century and contained in his book, *Kalapuya Texts*. The society's culture and past are described through these stories, ranging from how to cook camas to how the Kalapuya came to be. Interviews with David Lewis, the cultural director at the Grand Ronde, help to provide the context to understand these stories. Exploration of these stories helps to reveal how Kalapuyan views of their relationship to the land, and how changes in the landscape changed

22 Nickolas Lormand with Larissa DeHaas

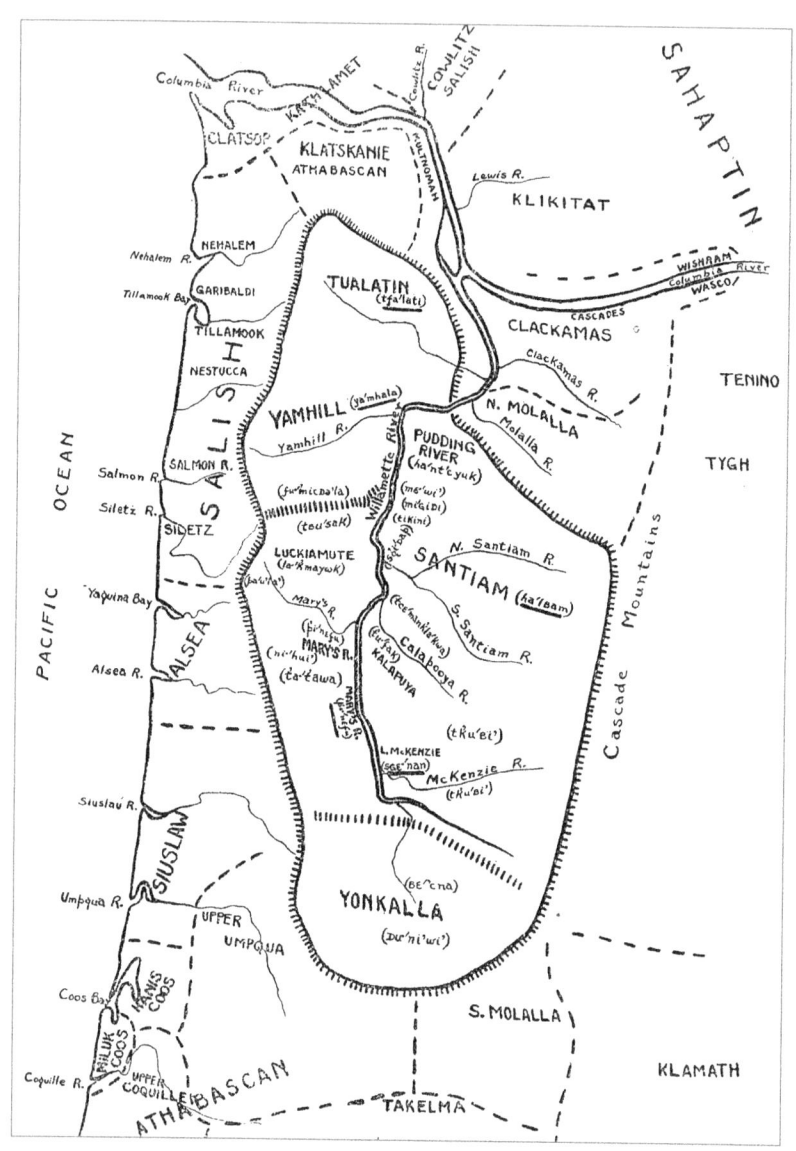

5. Map of Kalapuya territories, as reconstructed by Melville Jacobs. *Melville Jacobs,* **Kalapuya Texts** *(1945)*

cultural values. The stories tell the history of the dynamic relationship that humans have with the landscape around them.[2]

A Time for Stories

> The people used to say, "It is not good to tell myths in the summertime. Perhaps a rattlesnake might bite a person, or a yellow jacket might sting a person, should one tell myths in the summertime." But they do tell stories during wintertime. It is good to tell myths in the wintertime. There are long nights in wintertime.

This first story quickly illustrates the significance of the nonhuman natural world in Kalapuyan narrative. Here, the seasons shape when storytelling takes place. The land of Zena is prosperous and full of both human and natural activity during the summer. There are many dangerous animals that have to be avoided, such as a rattlesnake. But in winter, as people came back together from their semi-nomadic summer lives, the Kalapuyan community took time to catch up and connect around the fire. And they came together to share the burden of that season, for the winter could be very difficult throughout the Eola Hills, as another story explains:

> That moon, the people said, that moon some of the people ate their moccasins. It is an extremely bad moon. When that moon went by, and the next moon was indeed approaching now, then grouse sand. Now then they addressed the moon. They said, "We are indeed still here. Indeed now we have been dying in body." Old people addressed the moon. And then when these grouse sand, that was the time then when the snow fell hard, now the people would say, "Oh this is just a mere nothing. It is grouse's spirit-power-song, it is that sort of snow. It is in that manner that there is snow." That is the way people would speak. "It is because of the spirit-power-song of grouse that it is like this."

This story highlights the hardship that winter placed upon the Kalapuya, and provides an explanation for those seasonal challenges of interacting with the nonhuman natural world.[3]

Oral narratives help explain other types of environmental changes and catastrophic events. Through usage of monsters, large movements of societies, and animal characters, oral narratives provide captivating stories that teach history and explain why and how the landscape has changed over time. The Kalapuyan narrative, "Mosquito and Thunder," explains the phenomenon of lightning strikes this way:

> Mosquito was always telling it to thunder, when the thunder said to him, "Where have you gotten this blood?" Then mosquito would say, "Oh I get it from this white fir tree. That is where I get this blood. There is a lot of blood in this white fir tree."

This story is an example of the Kalapuya using narratives to provide an explanation for an otherwise unknown phenomena. The explanation for lightening is that mosquito lies to thunder and tells him he sucks blood from the white oak tree; hence, lightning always strikes these trees.

This story also provides a clear example of lessons learned from the nonhuman natural world, specifically, the lesson of thunder creating fire in the natural landscape, which the Kalapuya noticed and used for their own purposes to encourage the growth of particular foods and foliage for game. Walking up to the oak savanna at Zena, one can see camas growing on the right and clover throughout the field on the left. The oak trees propagate the field due to the diligence of recent burning. As shown in the narrative, the practice of burning by the Kalapuya has shaped the oak savanna landscape to be considered native. The Kalapuyan burning tradition is protected through conservation easements that now necessitate burning to preserve the oak savanna. Students and professors work side by side replanting native plants and ripping out the invasive species every year, largely unaware of the long-standing cultural context that has shaped what they plant.

Origins and Explanations

Every culture has a creation story. The Kalapuyan creation story, *The Four Myths Ages*, encompasses the larger history of the Kalapuya, while simultaneously introducing core cultural values. Making use of both physical references and symbols—such as bow and arrow, which are generally associated with the powers of life and death—the story moves through four ages (also translated as generations). What follows are shortened and summarized portions of that story, interspersed with analysis of what those stories reveal about the intersection between Kalapuyan history and the landscape.

> Long ago [in the myth age] there were people. There were many people, they filled this country. All the children who were made [born] became big [all grew up—there was no death in that age]. So then they accumulated for a long time. Now then five persons who were

hunters went away, one dog accompanied them. Now they slept five times. When it became dark the dog left. Then one small girl [back at home] spoke thus to the dog, How many [deer] have been killed [by the hunters]?" The dog did not speak. She spoke to it five times in that manner, and then the dog spoke thus, "Five [deer] have been killed." And now the earth turned over. All the people [of the first myth age] changed into stars [and they are still stars today].

Here the narrative begins with a large physical landscape change, and the change is the earth turning over. The number five is symbolic for the Kalapuya and is the indicator within this passage of a change coming. The widespread change reflects how large environmental changes could alter both the landscape and the amount of people that still existed after the impact. The environment forces a response from the people inhabiting the area. Here, this response is that the Kalapuyan ancestors "die" and then become part of both nature and the creation. This profoundly and literally connects the culture to nature. The Kalapuya created the stars in the sky, and when their descendents look up and remember this part of creation, they know their ancestors are watching over them, and can feel the connection.

> Now one man spoke thus, "A great many persons who are nearby will arrive here, those who are the new people. It is better than we be no more in this country." The headman went all over. Now then all those people were changed into stones. Here in the water (today still) are quantities of such small pebbles. Long ago those were the people.

This section further shows the Kalapuya becoming a part of the larger cosmos. Here is a direct reference to the environmental components that were a part of the historical landscape. The passage of time translates through the changing of stone to pebbles. There are small pebbles throughout the Willamette Valley landscape, found in the river layer in the stratigraphy. The movement of the water shapes larger rocks into small, round, smooth pebbles that are a part of older rivers. The changing water system throughout time creates distinct layers in the stratigraphy. This narrative not only ties the Kalapuyan ancestors into another section of creation, it also shows the extensive history of the Kalapuya throughout the Eola Hills. The Kalapuya came to understand their culture as a timeless feature of the valley. Their "sense of place" stretches deep into the sands of time before the layers of stratigraphy were even laid down.

> Long ago there was no water. Water was only pulled from trees. Now all (sorts of) of people were again on the earth. The third (series or generation of myth age) became many. Now then two women stole one infant, and they kept it all the time. It became large and dug roots. It turned into a girl. Then one flint boy found her, and he brought her to her mother. Then the two women became angry, they stood and danced (at their spirit power dancing), they made rain and it rained twenty days. The earth was completely full, the mountains sank, and then the people died. Only one boy and one girl were left on the earth. Then the water went back. He saw those two women who made the water, and so then he killed them. Flint man burned those two women, he took their ashes and blew them skywards. He made fog, mist and clouds. This is what he said, "You are not to go on the earth, and you are to be clouds forever now. When the clouds become thick it will rain." All of the persons in the flood turned into beaver, changed to steelhead, changed to all the kinds of things to live in the water. Formerly they were persons, from here on now they lived in the sea. Formerly they were our own people, the Water Being, steelhead, crawfish, salmon trout, mink, land otter, sea otter, seal, the spotted dogs of the whale, the various things of the water.

Multiple times, the ancestors of the Kalapuya became a part of the heavens, and the people were carried on through a girl and a dog. The blood of the animals runs through the veins of the second age of people. Here, respect for animals becomes an instilled cultural value. The Kalapuyan ancestors are seen as creating all ocean life. Respecting your elders in this sense means respecting all life forms. Once again the land is free from large amounts of water and the people are free to roam and live, now with more animals and instilled respect for the creation of them. This part of the narrative also provides an example of the Kalapuyan chronology. The water is stated as receding away and application of this knowledge could be connected to the time of the Missoula floods throughout the Willamette valley. This would date the existence of the Kalapuya back to at least 13,000 years ago. If so, many landscape changes have taken place, and the subsistence for the Kalapuya has also altered.

> Crow entered their house, it spoke thus: "Make an arrow, and make a bow!" Crow spoke to him like this: "Hunt in the woods! Kill deer!

Kill elk! Kill black bear! Kill panther! Kill wild cat! Kill grizzly! Kill that kind of things! Eat the flesh! Make blankets from its hide, all sorts of things from its hide. They are good to wear. Make yourselves wealthy people." That is what crow said. Now then he told the woman as follows, "Make (and) sharpen a stick, sharpen the end of a stick, and dig a hole. Get these camas, and get these potatoes and get wild carrots, all the edible things in the ground so that they may be eaten."

A faunal change within the Willamette Valley after the large flood events may have caused this adaption within the culture, leading to technological changes. It is explained using crow, an important character in many other narratives, who instructs the woman on how to make a better life by utilizing the land in a variety of ways. The water moved along, it went all over, the water was started. First it became the ocean, and then it made streams, and then it made all the creeks, and now all the various types of waters were finished. "Now I have completed all the waters." That was what he said. "Now the water is fine. When the new people (the Indians to come) have arrived, there will be lots of water. They will not be poor in water." And so we are still living here now. (This is) the end of the myth people.[4]

The young boy then decides that the next people coming to the area will most definitely need water. So he goes to the moon and then to the sun, and with the sun's daughter he goes into a canoe, and they travel with the canoe. The young boy continuously speaks to the water to come with them, to follow them, and to spread throughout the land to make it rich for the next people, the Kalapuya. He gives them the gift of water, the richest gift of all. The narrative emphasizes the creation of water and the water systems of the world. Water is immensely important to those living in the Willamette Valley both agriculturally and socially. The creation of the water systems is important because the large amount of water is what creates the temperate rainforest biome of the valley. It is abundant and plentiful, and provides everything the Kalapuya need. Water is the most important resource for life forms, and is the most abundant resource at Zena. It holds a special significance for the Kalapuya, and for every member of the Willamette Valley.

This creation story, in sum, shows the progression of geological time and the resulting widespread landscape changes. As time progresses along the geological time scale, four transitions narrate the creation of new Earth. Each of the "myth ages" ends with the land being readied for the next generation of Kalapuyan ancestors or gods to come and live fruitfully. A transition of the land is explained as one of the primary drivers, with assistance from Crow being common. From the narrative above, for example, the earth is flipped upside down and the Kalapuyan ancestors come to be the stars. The large flood initiated the people changing into sea animals, and then the creation of the all-encompassing water systems of the earth. The Willamette Valley is abundant with both lakes and rivers. Just as the stream that runs through Zena changes which plants can live in that area, rivers and streams are important movers throughout the Kalapuyan oral narratives. The Kalapuyan ancestors are literally a part of the landscape for the expansion of the people during each myth age transition. This detail hints at the deep respect the Kalapuya have for the landscape. Their ancestors are direct contributors to the creation of the world. It is important to respect the environment that was given to them. The subsistence changes with the shifts throughout the ages, as descendants learn to harvest camas and hunt animals. The intertwined narrative of animals, environment, and people shows the inter-complexities of the world surrounding those living in the Willamette valley. These stories show how the Kalapuya made sense of the world, and that their sense of the world created a cyclic, sustainable relationship.

Learning from Storytelling

The wisdom of oral narratives is in bringing an entire community together and sharing knowledge, which builds strong, meaningful and everlasting relationships. We all need a community to relate with, just as we all need water to survive. In this way, we are all related. Likewise, stories are our common language. During one's "coming of age" years, an individual is required to find where she or he fits within the community. It is a long process of learning and challenges and asking the bigger question: "Where is my place in this world?" It may not be as straightforward today as it was for the Kalapuya, who had deeply rooted history and strong societal requirements within the Willamette Valley. There is no specific time, like a spirit quest, when your parents tell you that it is time for you to go out and think deeply and seek spiritual guidance for what you are meant to do. Maybe that is why so many go to university and then graduate school and still feel lost without a purpose, or a place. So as a culture, are we losing our sense of place? Are globalization and the "melting pot" making the defining characteristics blurry? Is it getting harder to find a place amidst urban centers and a degraded environment? Have you asked yourself: where is my place in this world?

Notes

1. Basso, *Wisdom Sits in Places*; Lang, "From Where We Are Standing," 90.
2. Jacobs, *Kalapuya Texts*; this chapter summarizes and explains some of those texts, which are not always the complete transcribed version by Melville Jacobs. David Lewis, in discussion with Nickolas Lormand, 12 April 2012, and with Larissa DeHaas, 17 October 2012.
3. Mackey, *The Kalapuyans: A Sourcebook*, 44.
4. Jacobs, *Kalapuya Texts*, 173.

3

Kalapuyan Interactions with the Land

Brayton Noll with Summer Tucker

In the Kalapuyan narrative, "The Four Myth Ages," Crow instructs humans, saying "Make (and) sharpen a stick, sharpen the end of a stick, and dig a hole. . . . Get these camas, and get these potatoes, and get wild carrots (and) all the edible things in the ground so that they may be eaten. . . . (Then) you will be well off." This small excerpt suggests the myriad ways in which the Kalapuya interacted with the environment of the Willamette Valley, including, quite possibly, Zena. Cultivating and harvesting food was just one of the ways that the Kalapuya made use of the nonhuman natural world: they built dwellings, developed medicines, and engaged in burning practices, all in an effort to, in the words of Crow, "be well off." Through these interactions, the Kalapuya developed a deep understanding of their environment, and by investigating these interactions, we might better understand at least a part of the Kalapuyan sense of place.[1]

The theoretical framework of Traditional Ecological Knowledge (TEK) suggests some ways of understanding how the Kalapuya interacted with and made use of the nonhuman natural world. As the ecologist Fikre Berkes explains, TEK refers to "a cumulative body of knowledge, practice, and belief, evolving by adaptive processes and handed down through generations by cultural transmission, about the relationship of living beings (including humans) with one another and with their environment." More specifically, TEK develops through "careful and repeated personal observation," contextualization within a particular location, and incorporation of "spiritual/religious" ideas, along with morals. Most importantly, though, TEK comes from an environment in which its users have unique perspectives about the world. Additionally, as the philosopher Fulvio Mazzochi notes, TEK is "embedded within specific worldviews" and is something that cannot be entirely comprehended without having "a fuller understanding of the culture taken as a whole." TEK, then, is not simply a number of different methods for using natural resources, but a reflection of a culture's views of how to interact with the environment.[2]

To investigate both TEK practices in Kalapuyan life and the views that accompanied those practices, this chapter draws on recent surveys of the landscape and vegetation at Zena; an extensive interview with David Lewis, the Cultural Resource Manager for the Confederated Tribes of Grand Ronde; and the work of the historians Peter Boag, Robert Boyd, Judy Juntunen, May Dasch, and Ann Rogers. Taken as a whole, this research reveals some of the ways in which the Kalapuya interacted with the environment and suggests how those interactions helped shape their sense of place.

A Cyclical Life

In the background of the specific ways that the Kalapuya used plants, animals, and the landscape, there were some overarching concepts that guided Kalapuyan practices in relation to nature. One particular theme present in Kalapuyan life was a focus on a cyclical relationship with the natural world. The Kalapuyan calendar reflects this relationship: the names for each month represented the "annual growing cycle." For the Kalapuya, the names they had for each lunar month reflected certain activities of that time. For instance, April is the "budding month" while February is the "out of provision month." In practice, a cyclical relationship can also be found in the Kalapuyan practice of the seasonal round. David Lewis describes the seasonal round as a cycle where the Kalapuya would move from different geographic locations throughout the year. Around the beginning of the Kalapuyan year, September, the Kalapuya were in the late summer and early fall period of the seasonal round, and could be found settled in the Willamette Valley. Towards the end of winter, the Kalapuya would relocate, moving to the floodplains of the Willamette River. In this system of movement, the Kalapuya used different places for different purposes. For instance, when the Kalapuya were located in low, wet prairies, they would create pits that were used to prepare camas. In contrast, the dry prairie was where the Kalapuya would utilize and prepare seeds and nuts for consumption. As David Lewis explains, for the Kalapuya, the process of moving to and from place to place throughout the year was an "annual cycle." Traveling across the land, the Kalapuya not only created a cycle in interacting with the land, but also clearly had a distinct view of nature and their relationship with it in their conception of the year.[3]

Dwellings

David Lewis compares the cyclical nature of this Kalapuyan practice as analogous to the regular movements we now make between home and various places where we obtain products that we use in our lives. And so perhaps the most tangible way to learn about how the Kalapuya lived is to imagine a day in their shoes, so to speak. Usually, when one thinks about the beginning of the day, they would likely imagine themselves waking up in bed in their home. So, in what kind of dwelling would the Kalapuya have found themselves? If it was winter time, the Kalapuya lived in a "main village" structure. In general, these structures shared common features like walls, roofs, and usage of wood. However, it does seem that there was some variety in these dwellings. For instance, the Santiam band of the Kalapuya would dig partially into the ground in creating these homes, and thus, the ground would serve as a floor and as part of the walls. Alternatively, the Tualatin band of the Kalapuya was found to have chained separate homes together, ultimately forming a large overall home. When it comes to natural resources used to create these homes, it does appear that a variety of materials were used, perhaps due to the variation in how homes were constructed. Grass, for instance, was sometimes used in constructing walls, which is unsurprising given the grasslands to which the Kalapuya had access. At Zena today, when walking in the oak savanna areas of the property, one can easily end up traversing grasses along with the trees that the Kalapuya most likely in the Zena area (the Luckiamute or Yamhill bands) would have used for building structures. For instance, one can find a Ponderosa Pine standing in an oak savanna area, while further into the woods, Douglas fir are packed alongside the trail. Other trees the Kalapuya used include the Western red cedar, Oregon ash, and Grand fir, all of which have been identified as being present at Zena.[4]

Food

As a hunting and gathering society, the Kalapuya consumed a wide range of foods, a strategy that, as Peter Boag notes, "guaranteed security." However, plants were a particularly prominent part of the Kalapuyan diet. The Kalapuya used more than fifty plants,

6. **Acorns found at Zena.** *Courtesy of Larissa DeHaas*

including camas, tarweed, blackberries, cattail, and lomatium. If looking for a snack at Zena today, one would be able to spot blackberries alongside the road or further into the woods, easily available to pick and eat. One would also be able to locate camas in the "northern east-west drainage area" of Zena. Additionally, one might find a kind of tarweed, mountain tarweed, at Zena. The Kalapuya used both camas and tarweed extensively in their diet. A drainage area at Zena would be a particularly good growing spot for camas, which thrives in moist areas.[5]

In collecting camas, the Kalapuya would target the plant's bulbs and shoots. To prepare the camas, the Kalapuya used camas ovens, which they carved out of the ground. Evidence of these ovens existing over 12,000 years ago also points to a long term use of the plant. As scholars Judy Juntunen, May Dasch, and Ann Rogers have described, these ovens played an important role in processing camas.

> When the oven was full, they covered the camas with another layer of green grasses or leaves followed by a layer of dirt. The bulbs were baked for a day or two. Charred bulbs could be eaten immediately or laid out to dry in the sun. Cooks used a mortar and pestle to grind some of the dried bulbs into flour and pressed it into camas cakes about two to three inches thick and from three to six inches across. They stored them for use in winter, along with whole dried bulbs. They also made larger cakes that could weigh up to ten pounds and traded them for items they needed from other Indian groups.

In terms of the seasonal round, Kalapuya could be found gathering and preparing camas during the late spring and early summer.

Tarweed, another important plant in a Kalapuya's diet, was prized for its seeds. With tarweed, the Kalapuya would burn a resin off the plant to be able to better access the seeds. Like camas, tarweed would be subjected to fire, and this process would lead to transforming tarweed seeds into meal. Again, like camas, tarweed has a sweet flavor. For the Tualatin Kalapuya, tarweed was important enough that this band had different patches of tarweed that different people owned. As important plants to the Kalapuya, camas and tarweed are excellent examples of the knowledge the Kalapuya needed in order to be able to properly collect and prepare plant based food.[6]

Beyond consuming plants, the Kalapuya also spent time hunting. At Zena, Kalapuya may have been able to use some of the area's Oregon white oak to fashion a bow. As of 2005, these oaks were present in the southern portion of Zena, along with smatterings of oak in the eastern portion of the property, close to the area a person would enter Zena today from the road. Animals the Kalapuya consumed included birds, lamprey, salmon, black bears, and elk, among many others. Overall, looking at how they used plants and animals, it is clear that the Kalapuya had access to many different natural resources, and had well developed ways of using those resources.[7]

Tools and Medicine

Even though the Kalapuya were certainly concerned with utilizing plants for food, plants were also important for creating practical objects. For instance, baskets made out of plant materials were used for storing food. Baskets required a variety of plant materials, including bark, branches, leaves, and even shoots. Specific plants the Kalapuya used for baskets included Western red cedar, the Sitka spruce, and the California hazel. Creating a basket required proper harvesting techniques, preparation, and the knowledge to weave materials together into a useful item. If one were to attempt that process today, given some exploration, it is likely the person would find Western red cedar at Zena. Similarly, in making objects like bows, spears, or woodworking tools, the Kalapuya may have utilized other plants like the Pacific yew, Oregon white oak, serviceberry, and ocean spray. In the oak savanna environment that is present at Zena today, one would be able to find and utilize Oregon white oak for creating tools.[8]

The Kalapuya also treated sickness with a variety of plants, many of which are present at Zena today. With the details provided by scholars Judy Juntunen, May Dasch, and Ann Rogers, it is clear that plants were used in a variety of ways to deal with sickness.

7. **Yarrow found at Zena.** *Courtesy of Larissa DeHaas*

The Kalapuya used sword fern for a range of physical problems, from dandruff to birthing pains. Sword fern was applied directly to the body, chewed, or drunk as tea, and is present in copious quantities on a hillside at Zena. Tea made from yew needles targets internal pain. Yarrow, another medicinal plant, has also been found at Zena, and could be utilized in a variety forms, including as a "bath, . . . tea, . . . and poultice." Stinging nettle is a particularly versatile plant and might have been useful for an older person, as it addressed "rheumatism, . . . soreness, and stiffness." Some plants that the Kalapuya used as medicines were used for other purposes, as well. Not just used for construction, the smaller parts of the Western red cedar were used for medicinal purposes, including treatment for toothaches and as a gargle. Yew, yarrow, sword fern, stinging nettle, and the Western red cedar, are just some of the many plants used in Kalapuyan medicine.[9]

Burning the Land

As David Lewis explains, the practice of burning is "literally a rebirth. . . . You're renewing, . . . helping the land rebirth." The Kalapuya set burns to the land in the late summer and early fall, which are close the beginning of the Kalapuyan year in September. In burning the land, the Kalapuya significantly encouraged the environment to favor oak savanna. Beyond this impact on the natural environment though, burning was a way for the Kalapuya to enhance the environment for the plants and animals that were a part of their lives.[10]

The Kalapuya practiced burning in part to encourage the growth of particular plants, especially tarweed. Burning loosened tarweed seeds, making them more readily available. Along with making a plant more easily usable, burning could also increase yields of desirable parts of plants. For example, burning helped create higher yields of acorns and berries, along with making it easier to access acorns. Similarly, burning made the land more amenable for the growth of tobacco, liliaceous camas, bracken, wild onion, and lupine. Burning also caused certain plants like hazel and willow to grow straighter and stronger, which made them even more useful for basket making.[11]

Besides affecting plants, burning also facilitated hunting. The Kalapuya would use fire to push animals in a space surrounded by fire, a process known as *battue*. As those animals became increasingly surrounded by the battue, the Kalapuya would then hunt those animals (specifically, deer and elk). Another way that the Kalapuya used fire to ease hunting was to burn away underbrush that would otherwise obscure clear vision of animals to be hunted. Even more strategically, though, the Kalapuya likely used fire to maintain certain "microhabitats" preferred by desired prey like deer.[12]

Perceptions of the Nonhuman Natural World

Just as TEK includes knowledge about a specific environment, it also clearly involves attitudes and cultural perceptions. As David Lewis explains, the Kalapuya saw animals as "having spirits themselves. . . . And we try to speak to that." However, as Lewis notes, "we object to people who place us in natural history museums as being sort of nature. You know we're humans, so we're somewhat above the idea of the natural environment." Straddling an interesting divide between being distinct from yet also deeply connected to nature, the Kalapuya saw themselves as "stewards of the Willamette Valley." Lewis notes that "we need to respect that natural environment" because it is an entity that "gives us so much; [it] gives us the ability to survive."[13]

TEK and the Kalpuyan Sense of Place

It seems clear the Kalapuya utilized TEK. The practice of burning is one of the best examples of TEK in how it brings together many ways that the Kalapuya utilized and understood nature. Given the distinct impact on the environment in preserving oak savanna, knowledge of burning was clearly passed on from generation to generation among the Kalapuya and reflects "careful and repeated personal observation"—or, to put it in another way, science. Similarly, the practice of the seasonal round and the specific timing of activities seen in the Kalapuyan calendar is other evidence of how the Kalapuya had a clearly defined system of interacting with their landscape. Today, members of the Grand Ronde tribe still harvest berries and camas, indicating the longevity of Kalapuyan knowledge. And that knowledge left a lasting imprint on the land: consider Willamette University's current efforts to restore oak savannas, which were so often the result of deliberate Kalapuyan burning practices.

This examination of TEK interactions between the Kalapuya and nature reveals a few themes of the Kalapuyan sense of place. First, the Kalapuya had a very clear understanding of the world they lived in, and they knew how to use it to their own benefit, as demonstrated by their use of camas, tarweed, and especially burning. However, the natural world was not simply a tool for the Kalapuya but an entity deserving of respect. Perhaps, then, part of the Kalapuyan sense of place was not just that nature was something that they understood and influenced. Rather, as David Lewis's summarizes, "it's not just us." The Kalapuyan sense of place was not only grounded in the concerns of humans, but reached into concerns about a wider world of humanity along with the many parts of nature.[14]

Notes

1. Jacobs, *Kalapuya Texts*, 176.

2. Usher, "Traditional Ecological Knowledge in Environmental Assessment and Management," 188; Burkes, *Sacred Ecology*, 11; Inglis, *Traditional Ecological Knowledge: Concepts and Cases*; Mazzocchi, "Analyzing Knowledge as Part of a Cultural Framework," 45; Rinkevich et al., "Traditional Ecological Knowledge for Application by Service Scientists."

3. Boag, "The Calapooian Matrix," 24, 25, 39; Juntunen et al., *The World of the Kalapuya*, 23, 24; David Lewis, in discussion with Summer Tucker, 17 October 2012; Boyd, *Indians, Fire, and the Land*, 123.

4. Lewis, in discussion with Tucker; Juntunen, et al., *The World of the Kalapuya*, 14, 30–33;

5. Collins, *The Cultural Position of the Kalapuya*, 39; Boag, "The Calapooian Matrix," 25; Juntunen et al., *The World of the Kalapuya*, 48–53; Salix Associates, "The Zena Property," 10–12; Boyd, *Indians, Fire, and the Land*, 113; Stevens, "Common Camas."

6. Juntunen et al., *The World of the Kalapuya*, 48–49, 59; Boyd, *Indians, Fire, and the Land*, 114, 127.

7. Juntunen et al., *The World of the Kalapuya*, 41–42, 59–61; Salix Associates, "The Zena Property," 9; Savoca, "Land Use and Vegetation Change," 18.

8. Juntunen et al., *The World of the Kalapuya*, 37–42; Salix Associates, "The Zena Property," 9; Sims, "Forest Stewardship Plan," 5.

9. Juntunen et al., *The World of the Kalapuya*, 64–67; Salix Associates, "The Zena Property," 3, 10.

10. Lewis, in discussion with Tucker; Boyd, *Indians, Fire, and the Land*, 122; Juntunen et al., *The World of the Kalapuya*, 23–25.

11. Boyd, *Indians, Fire, and the Land*, 114, 117–22; Boag, "The Calapooian Matrix," 29; Lewis, in discussion with Tucker.

12. Boag, "The Calapooian Matrix," 30; Boyd, *Indians, Fire, and the Land*, 113, 118; Juntunen et al., *The World of the Kalapuya*, 26.

13. Berkes, *Sacred Ecology*, 11; Mazzocchi, "Analyzing Knowledge as Part of a Cultural Framework," 45; Lewis, in discussion with Tucker.

14. Lewis, in discussion with Tucker; Juntunen et al., *The World of the Kalapuya*, 24.

Kalapuyan Society and Sense of Place at Zena

Elena Crecelius

The society of the Kalapuyan people reflected their sense of place, and the Kalapuya influenced the land in certain ways because of their social structure. The Kalapuyan sense of place was greatly influenced by spatiotemporal divisions, social structure, and gender dynamics. The interaction between the Kalapuya and the land at Zena led to the development of an interlinked sense of place and society. In fact, the depth of that relationship, when threatened, facilitated their near annihilation and ultimate removal from the land.

Spatiotemporal Divisions

Spatial and temporal divisions occurred at a yearly, seasonal, and monthly scale within Kalapuyan society. In the warm months the Kalapuya traveled through the foothills of the Willamette Valley, and to Zena, on their subsistence rounds. Although they only spent a few months of the year on Zena, the connection to and understanding of the land was deep because of their close interaction with the nonhuman natural world. While at Zena the Kalapuya lived in temporary shelters that usually consisted of tree branch homes under trees. The climatic factors at Zena greatly impacted the Kalapuyan interactions with the land during these subsistence rounds since they had minimal shelter from the elements. The Kalapuya worked directly on the land during these summer months: finding and harvesting the resources, hunting and fishing, weaving baskets to transport their harvest, and preparing the harvest for transport. Another temporal division is the seasonality of burning, which occurred in late summer and early fall and would have required the Kalapuya to visit Zena. The Kalapuya named their twelve lunar months for natural events that took place at that time. For example, August was designated as "camas time" and was the time to dig camas. October was the month when the "hair [leaves] fall off" and was the time for the Kalapuya to harvest *wapato*. The natural monthly events were so important to the Kalapuya that they were the primary way the Kalapuya identified temporal divisions. These spatiotemporal divisions were integral to the development and maintenance of the social Kalapuyan structure and highly influential on the amount of time that the Kalapuya spent at Zena. These divisions along the lines of space and time were used to support the Kalapuyan social structure and greatly affected the interaction of the people with the land.[1]

Social Structures

Kalapuyan society at the level of a band—including the Luckiamute, which travelled to the Zena area during warm months—was made up of different social classes that were determined by material wealth and spirit-power. At the top of the social structure were the chiefs, who were likely the most influential when decisions were made concerning the land. The election of the chiefs was based on honesty, acquaintance, and wealth; these factors were important because a chief had to be able to care for and assist members of her or his band. The importance of providing for the band in determining leadership speaks to the importance of the land to the Kalapuyan leaders, since they would want to maximize harvests while ensuring future resource availability. Therefore, the land at Zena would have been important to the chiefs in terms of land management and supervising the subsistence rounds. While the requirements did not necessarily mean that leadership passed through bloodlines, it was common for chieftainship to pass through a family because of the passage of the relevant factors from parent to child. This would result in the same family being largely in charge of decisions made about the land. However, chiefs were not alone in making decisions; they participated in a council that included other influential people, such as the shamans.[2]

Though usually lacking material wealth, shamans possessed significant power, due to their centrality to spiritual and cultural ceremonies that determined most decisions and processes within Kalapuyan society. If the natural world somehow failed the Kalapuya people, it was the shaman who would dance and have visions to determine the cause. Shamans conducted spiritual ceremonies such as those before the first harvests and in treatment of disease. Particular parts of the nonhuman natural world were central to the shaman's ceremonies; for example, water was used to cleanse, and plants were used in most ceremonies and treatments. The shamans' activities speak to the intermingling of the natural and social worlds of the Kalapuya and the importance of the nonhuman natural world in the development of their sense of place.[3]

The social levels beneath the leaders were also determined by wealth and power, and those different cultural standings and subsequent interactions with different parts of the environment would have produced different senses of place. Slaves provide an interesting subject for analysis. The slaves at the bottom of the social structure were usually obtained as prisoners of war from other tribes; however, orphans within a band were also demoted to slave status. Slaves lived with the family who owned them and spent a large amount of time working on the land doing chores and gender-specific tasks. However, while the slaves may have been proximate to the Kalapuyan oral traditions and beliefs, they were probably not included in ceremonies and myth-telling. Kalapuyan slaves would therefore have a strong connection with the land but their sense of place would have been distinct from that of people at higher societal levels.[4]

While material wealth could be gained through family connections or trade, spirit-power was directly derived from interactions with the nonhuman natural world and was present in all aspects of Kalapuyan society. The Kalapuya believed that, "all of the forests, fields, and shore were alive with spirit powers," as were the Kalapuya themselves. An individual's spirit-power could be increased or diminished by his or her interactions with the nonhuman natural world. Spirit-power was largely attained on a five-day solo journey taken by Kalapuya youth into the wilderness. Individuals could also publicly display or prove their spirit-power. For example, while hunting grizzly bears a man could hold the pole to poke the bear and distract it while the others shot it with arrows. If the man was successful at keeping the bear at bay with the pole, it was considered a display of strength and courage and indicated strong spirit-power. In these and other ways, interactions with the land were crucial to shaping both an individual's connection to the nonhuman natural world and that individual's place within Kalapuyan society.[5]

Gender Divisions

Knowledge of the nonhuman natural world gained through labor helps to deepen the understanding of that place. In fact, knowledge of the land provides intellectual, economic, and spiritual connections to land. With this heightened knowledge come more possibilities to interact with the land through land management practices, resource extraction, and developing cultural and spiritual practices rooted directly in the land. Every interaction with the land provided the opportunity to deepen the sense of that place and increased the importance of that place to the society and individual. Division of labor along gender lines would have resulted in different place-based knowledge between women and men because of interaction with different parts of the land at Zena.[6]

Although women could be chiefs or shamans, they were not allowed to be part of councils, and therefore were excluded from some decisions about land use and management. But Kalapuyan women shaped the landscape in a variety of ways. Kalapuyan women at all social levels were responsible for gathering food from the land. The women would have lived at Zena during the subsistence rounds. Women also made and carried the baskets used for transporting the harvested resources from their subsistence rounds. To take another example of how women shaped the landscape, Kalapuyan burning practices were not only used to keep land open for hunting but also to increase the yield of plants harvested by women. These interactions with the landscape, in turn, produced intense first-hand knowledge of the land.[7]

Kalapuyan men at all social levels undertook the hunting, fishing, and manufacturing of tools. The men also were in charge of burning the land to facilitate hunting and harvesting of plants. The burning patterns depended on a "yearly round of subsistence activities and seasonal pattern of plant succession" and were another example of the deep knowledge of the nonhuman natural world. These gender divisions, as well as larger social structures, resulted in a reciprocal relationship between sense of place and Kalapuyan society. It was this place-based society that the Euro-Americans encountered upon their arrival at the beginning of the nineteenth century.[8]

Euro-American Arrival and Disease

When the American explorers Meriwether Lewis and William Clark passed through the area in 1805 and 1806, they listed the Kalapuyan population at 9,000 people, based on their Chinook translators' estimates. The first recorded contact between Kalapuya and Euro-Americans came in 1812 when explorers and the Pacific Fur Trading Company traveled through the Willamette Valley. Another fur trading company, the Hudson Bay Company, estimated the Kalapuyan population at 7,785 to 8,780 people in the Willamette Valley in the 1820s. An early nineteenth-century Kalapuyan population of 8,200 is considered most accurate by modern-day historians. These population estimates include all the bands of Kalapuya within the Willamette Valley. Despite the presence of this population, Euro-Americans often did not see—or would not recognize—evidence of human manipulation of the landscape, and so they believed the land to be essentially unoccupied and unused wilderness. This misconception greatly affected the relationship between Euro-Americans and Kalapuya but was in no way the greatest influence on the Kalapuya after the Euro-American influx in the nineteenth century.[9]

The most influential things brought by Euro-Americans to Kalapuyan society were diseases, especially smallpox, fevers, and malaria. Between 1782 and 1783 a wave of smallpox swept through the Kalapuyan bands in the Willamette Valley. This smallpox originated from the Midwest and killed half the native peoples of the Pacific Northwest. It was estimated that the Kalapuya lost about 2,000 people to smallpox. Fever reigned among the Kalapuyan people between 1830 and 1833; its source was most likely the trading and exploratory ships that traveled along the Columbia River. The disease likely traveled from the native peoples of the Columbia River Basin down into the Willamette Valley through trade and cultural interactions. Along the Willamette River 5,000 to 6,000 Kalapuyan people were reported dead due to this fever epidemic. Despite these preliminary horrific losses to disease, the greatest losses in the Willamette Valley came from the malarial plague that reoccurred in the summers of 1830 to 1833. Malaria is considered the "most influential disease in the Willamette Valley" among historians studying the transition of settlement between the Kalapuya and the Euro-Americans. In fact, this disease "probably constituted the single most important epidemiological event in the recorded history of what would eventually become the state of Oregon."[10]

The connection between the Kalapuya and their sense of place at Zena exacerbated the effects of disease. A variety of Kalapuyan sociocultural factors and practices increased disease virulence and communication. The practice of communal living facilitated the spread of disease among members of the family and band. When women made baskets to carry the harvested resources, they wetted the plant fibers in their mouths, making the baskets more pliable yet increasing the presence of disease-causing microorganisms. Chances of infection were increased by the lack of windows in the Kalapuya's permanent homes, which restricted air flow, and dirt floors were impossible to sanitize. Traditional treatments, such as sweat lodges and denying the patient water to drink, only exacerbated these foreign diseases. The yearly subsistence rounds of the Kalapuya took them through the most dangerous areas of the valley: wet lowlands that provided prime breeding grounds for malarial vectors. The trade patterns through which the Kalapuya obtained goods from other lands served to increase interpersonal contact and share disease along with goods. The immense impact that the Kalapuyan societal structure had on their ability to fight off the disease shows the importance of their social practices developed in conjunction with their sense of place.[11]

Extreme population loss due to disease reverberated throughout Kalapuyan society. Diseases directly resulted in a population loss of about 80 percent. But the epidemics produced other indirect losses and casualties as well. Kalapuyan women experienced depressed fertility rates that resulted in lower birthrates even when the disease was inactive. Marriages were disrupted after the epidemics, also resulting in lower birthrates. The most abstract, yet perhaps most influential, result of the epidemics was the cultural crisis of the Kalapuya and the loss of confidence in their society at the individual level. The decreased population diminished the society's ability to maintain the structures required for task groups such as the subsistence rounds and leadership councils. Due to the oral nature of Kalapuyan society, the population loss led to depletion of cultural resources and traditional knowledge. Such damages brought on by disease weakened the Kalapuyan tie to the land by disrupting their cultural practices, social structure, and interactions with the land.[12]

Conclusion

The Kalapuyan interaction with the nonhuman natural world was integral to the development of their culture as well as the destruction of their way of life. Kalapuyan social structures and gender dynamics, which had developed from interactions with their land, played major roles in facilitating the spread of disease. Subsequent population losses facilitated the increasing presence and expansion of Euro-Americans, who pushed the Kalapuya off their land and later restricted access to that land through the treaty process. This removal ended direct Kalapuyan interactions with the nonhuman natural world in the Zena area, and dramatically changed the ways in which they developed their sense of place at Zena.

Notes

1. Berg, *The First Oregonians*, 126; Mackey, *The Kalapuyans: A Sourcebook*, 40; Juntunen et al., *The World of the Kalapuya*, 23.

2. The number of chiefs varied between bands. There were sometimes up to three chiefs per village: one who remained in the permanent village to maintain communication, and others who acted as go-betweens who visited other bands and supervised the subsistence rounds; see Juntunen et al., *The World of the Kalapuya*, 17; Mackey, *The Kalapuyans: A Sourcebook*, 36.

3. Juntunen et al., *The World of the Kalapuya*, 19, 64–67; Collins, *The Cultural Position of the Kalapuya*, 52.

4. Juntunen et al., *The World of the Kalapuya*, 19.

5. Beckham, *The Indians of Western Oregon*, 85; Jacobs, *Kalapuya Texts*, 21–23.

6. Aikens et al., *Oregon Archaeology*, 397.

7. Mackey, *The Kalapuyans: A Sourcebook*, 39; Aikens et al., *Oregon Archaeology*, 409.

8. Boyd, "Strategies of Indian Burning in the Willamette Valley," in *Indians, Fire, and the Land*, 122.

9. Boyd, *Indians, Fire, and the Land*, 99, 238; Boyd, *Indians, Fire and the Land*, 99; Confederated Tribes of Grand Ronde, "The Kalapuya: A Wealthy Way of Life"; Boyd, *The Coming of the Spirit of Pestilence*, 241; Collins, *The Cultural Position of the Kalapuya*, Appendix; White, *Land Use, Environment, and Social Change*, 25.

10. Mackey, *The Kalapuyans: A Sourcebook*, 20–21; Boag, "The Calapooian Matrix"; White, *Land Use, Environment, and Social Change*, 27; Boyd, *The Coming of the Spirit of Pestilence*, 84.

11. Boag, "The Calapooian Matrix," 37; White, *Land Use, Environment, and Social Change*, 28.

12. Boyd, *The Coming of the Spirit of Pestilence*, 243, 269, 272, 275; White, *Land Use, Environment, and Social Change*, 29.

5

The Legal Development of Zena

Michael Harder with Alec Weeks

In 1865, Sanford Watson was legally granted ownership of a piece of land that is now part of Zena. Twelve years later, Edward Robson became the legal owner of a different parcel of land, a portion of which is now part of Zena. The final piece of Willamette's Zena consists of part of a 160-acre claim given to Thomas Warriner after his service in the Cayuse War, which started in 1847 and lasted till 1855. These three events mark a crucial change in landscape, when individuals first took sole and complete ownership of bounded land. Understanding how this came to be requires investigation of Euro-American laws and actions.

In his book *The Willamette Valley: Migration and Settlement on the Oregon Frontier*, William Bowen recognizes that legal mechanisms, particularly as expressed in maps, were important to the process of migration and settlement. He explains, "Location is the central theme, its description and explanation. The map is the primary form of expression." Furthering Bowen's work, this chapter explains both the development of the Euro-American boundaries that outline Zena and how these boundaries were crucial to the Euro-Americans' ability to conceptualize Zena. The legal mechanisms explored in this chapter—international agreements, ad hoc rules, legal precedent, and, finally, treaties with Native Americans—stemmed from and shaped a much deeper belief that a place must be bounded and owned before anyone can develop a meaningful sense of that place.[1]

The Northwest Ordinance

Westward expansion to areas like Zena began with a Euro-American vision that land was wild, untamed, and brimming with potential. From the earliest days of the United States, the federal government sought to bring order and the potential for cultivation and "civilization" to this lawless land. Congress first outlined the rules and procedures concerning interactions with land—and the native peoples—in the Northwest Ordinance of 1787. The ordinance established procedures by which "wild" land would be officially incorporated into the Union, a process that, ostensibly, began with the relationship between Native Americans and the federal government. The Northwest Ordinance promised that "the utmost good faith shall always be observed towards the Indians," including the protection of property and lands and a guarantee from invasion or from being disturbed.[2]

But that promise was often betrayed in practice, as Euro-Americans rushed to lay claim to and bound Native American land. The original idea of complete cadastral surveying—surveying of the boundaries of land parcels—of the West came from the Land Ordinance of 1785. The Northwest Ordinance made this feasible through the creation of townships: square plots of land six miles on each side were divided further into one-square-mile plots, and then again into 160-acre plots, called quarter sections. This checkerboarding of lands divided up much of what would become the United States, including, eventually, the area now known as Zena. The plotting of the land allowed Euro-American settlers to understand space. Without such rectilinear boundaries, Zena was nothing but a vast landscape, but with survey markers and maps, Zena transformed into a place that could be conceptualized and possessed.[3]

Oregon Treaty of 1846

Prior to getting an official United States survey, Zena had to officially become part of the United States. At the beginning of the nineteenth century, both Great Britain and the United States explored and became increasingly interested and invested in what came to be called "Oregon Country." After the War of 1812, Great Britain and the United States agreed to co-occupy the area, but by the 1840s it became apparent that Oregon Country was becoming predominately occupied by Americans. Although President James Polk blustered about going to war with the English to secure a border at the 54th parallel, Great Britain and the USA agreed in 1846 that the 49th parallel would be the border between the two nations.

The Oregon Treaty brought Zena within the legal jurisdiction of the United States, and established the basic legal framework of boundary, possession, and an official government. It also reserved for the United States the exclusive right to negotiate treaties with Native Americans throughout the Oregon Country. Technically, without such agreements, US citizens could not lay legal claim to land. But such technicalities were often ignored in the rush to the Willamette Valley.

The Oregon Territory

Zena, as elsewhere in the Willamette Valley, became populated with US citizens well before the federal government secured treaties and possession of Native American land. The initial settlers during this time could be considered squatters, as they were using the land for their own purposes before a legal title was passed. This was the case on Zena with Sanford Watson and his family, who lived on the land prior to 1852—before the United States had concluded treaty negotiations with the Kalapuya. Squatters like Watson staked out lands before the General Land Office (GLO), a federal agency in the Department of Interior, surveyed them or before the land was available for sale by the government. Sanford and others did so with the blessing of the Provisional Government of Oregon, an ad hoc government created in 1843 in anticipation of incorporation of Oregon Country into the United States. Under Article 3 of the Provisional Government's Organic Code, an individual could not exceed ownership of a one-square-mile section (640 acres) per person and had to make improvements on the land in six months. This would become the precedent for future land division.[4]

Oregon did not officially become a Territory until 1848. Violent conflict with the Native Americans contributed to this process. In 1847, after the death by disease of many Cayuse Indians on the Columbia Plateau, a group of Cayuse took action, killing fourteen whites, including the missionaries Marcus and Narcissa Whitman, and taking others captive. This sparked the Cayuse War, during which hundreds of US citizens from Oregon Country volunteered, and the provisional government demanded territorial recognition, which was granted in August 1848. Both the war and territorial status had profound effects on Zena, from the ownership to the mechanisms of ownership. Thomas Warriner served in the Cayuse War, and from his service was awarded a 160-acre plot. The Territorial Act of 1848 set up the basic legal framework of the area.

This framework was crucial, as it established a federal government presence and allowed for the division of the land and the creation of the plots that are Zena.

Oregon Donation Land Law

The Oregon Donation Land Claim Act (DLC) passed in 1850 established the procedures for the public to claim lands in the Oregon Territory. It placed the GLO in charge of, "administering, surveying, and [initiating] the disposition of the public domain lands." John B. Preston, the first surveyor general in Oregon, opened the survey office in 1851 in Oregon City. His surveying marked the start of the Willamette Meridian Rectangular Survey. Preston's work left profound effects on the Willamette Valley and Zena. Preston's legacy is the grid pattern that set up the structure for Zena's boundaries.

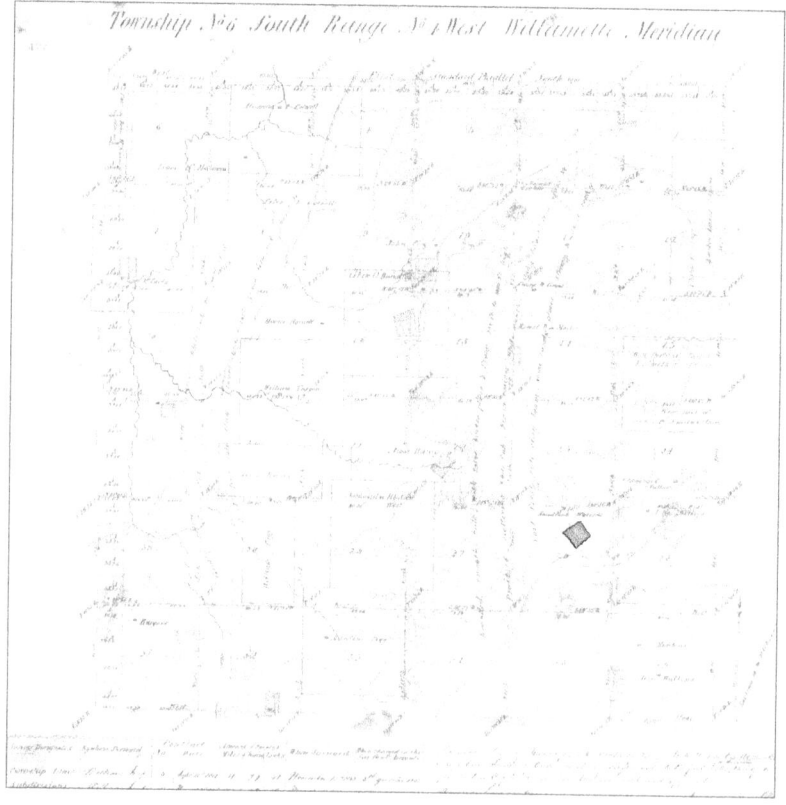

8. 1852 surveyor's map showing Zena-area township. Sanford Watson's claim is highlighted. *Courtesy of the Bureau of Land Management*

The DLC solved the issue of the Watson family's property ownership; Sanford Watson eventually received title in 1865. The Act legalized land ownership of people who squatted on the land and enabled new settlers, like Edward Robson, to go through a claim process. Before the DLC expired in 1855, claims were filed for almost 2.8 million acres of Oregon lands. Over 7,437 settlers filed claims on land. The DLC made the move west more popular, especially to the Willamette Valley. One estimate suggests that 25,000 to 30,000 immigrants entered Oregon, around a 300 percent increase.[5]

A primary factor leading to the implementation of the DLC was the growing population and the issue of remaining Native American populations. While the legislation involving the implementation of ideal land division onto the West reinforced this, social trends that started the movement were outpacing the federal government. As settlers were being drawn to the West, more pressure was applied on the government to address the growing conflicts with the Native Americans. Addressing these issues was crucial for establishing the needed legal mechanisms for settlers to gain ownership of, and, consequently, a sense of place at Zena.

Treaties

The treaty process with the Native Americans in the Willamette Valley began in 1850. The Willamette Valley Treaty Commission was in charge of ceding Indian titles to the valley and also ensuring the removal of the Kalapuya Indians and other tribes inhabiting the valley. This made possible the settlement of areas like Zena. The Commission worked to make treaties with several different tribes, but Congress decided to stop efforts almost a year after its creation in 1851. Unaware of Congress's actions, the Commission continued to work towards crucial agreements for securing Zena and the Willamette Valley. They ended up with six treaties, but none of these accomplished what the Commission was originally designed to do: the removal of all Native Americans from the Willamette Valley. The treaties were primarily promises of good faith between Indians and the Commission. Supplies and services like education, vocational services, health services, and protection from American settlers were given to the Indians in return for land rights. These treaties were not ratified in Congress and failed to remove Indians from the Valley. The population of the Willamette Valley continued to expand despite the lack of treaties with natives. The native people were promised a permanent reservation, the Coastal Reservation, which originally was about one third of the Oregon coast. These treaty conditions were never finalized.[6]

It was not until Joe Palmer became the Superintendent of Indian Affairs in 1853 that removal of the Native Americans gained momentum. Palmer aggressively pursued the creation of treaties, by whatever means he saw necessary. After acquiring signatures from Native Americans, he obtained amendments to the treaties in Congress and had them signed a second time by tribal leaders. Palmer gained some success through this second round of treaty making. This success had profound effects on the Willamette Valley and Zena, as it legally recognized the federal government as the owner or distributor of the land.[7]

On January 22, 1855, Palmer secured treaty agreements from the Confederate Bands of the Kalapuya. This event marked the official transfer of Zena from the Kalapuya to the hands of the federal government and the settlers. Joe Palmer's treaty offered 200,000 dollars in return for the title of 7.5 million acres of land, and was ratified in March 1855. The Confederate Bands of the Kalapuya Indians were forced onto the Grand Ronde Encampment during the winter of 1855 to 1856. The Reservation was formally created June 30 of 1857. Resistors to the waves of settlers, who entered and settled on the land well before the Treaty was signed, were finally removed and relocated to the South Yamhill Valley.[8]

The extension of the Northwest Ordinance into Oregon Territory through the DLC turned the lands into a transferable commodity. The transfer of land from the Native Americans also represented the transfer of a Euro-American sense of place into the Western lands. Settlers like Watson in Zena, who moved in before any formal agreements were made, suggest that the true intentions of the settlers and the government were to gain sole ownership rather than reach agreement with the natives. This suggests that the Euro-Americans were unable or unwilling to recognize the natives' presence due to the lack of "legal" possession by Native Americans.

Possessing Zena

With the Native American legal claim to Zena technically extinguished, the process of bounding and possessing Zena continued. Sanford Watson purchased a section of 640 acres in 1865. This includes the portion of the Zena property located in section 26. After his death, his land passed on to his four children and wife, who each received separate portions. The second owner was Edward Robson, who acquired his section in 1877, Donation Certificate #5022. His DLC, like all others, affirms that he "personally resided upon and cultivated" the land. The last quarter section, a 160-acre plot, was given to Thomas Warriner after his service in the Cayuse War. Warriner's property was sold to Daniel J. Fry after Warriner passed away in 1889. Daniel J. Fry "acquired Watson's and Robson's pieces, respectively, and from that point forward the 305 acres (while technically still two separate tax lots) transferred hands as one property." The property would then be passed on through families as the 305-acre section Willamette owns today. In short, after passage of the Donation Land Claim Act, it took 59 years for Zena to become delineated in the way it is today.[9]

A Bounded Sense of Place

As Zena became bounded and possessed, it transformed into an area that could be described and understood on a map. This allowed settlers to realize their relation with the land and the area around Zena. Both settlers and the government transformed the land in an effort to understand the land in a manner that was familiar. Laws acted to create technical relationships, boundaries, and to complete transactions. The law was used to reinforce Euro-American ideals, and in essence created the kind of place that Euro Americans could comprehend—part of the process of developing their particular sense of that particular place. The creation of boundaries illustrates how Euro-Americans embraced a particular conception of permanency. Titles were a crucial part of "sensing places" as it allowed all types of land to be understood, despite geographic differences, by creating lines that gained meaning through purchase and ownership. This system is disconnected from the environment, making it possible for an area to be understood in a broad sense. This broad sense of understanding is in essence the Euro-American sense of place: a less intimate connection, but a universal one.

Notes

1. Bowen, *The Willamette Valley: Migration and Settlement*, 5.
2. White, *"It's Your Misfortune and None of My Own,"* 137–55.
3. National Atlas of the United States, "What is the PLSS?"
4. Gray, *A History of Oregon, 1792–1849*; Robbins, *Landscapes of Promise*, 80–82.

5. Robbins, *Landscapes of Promise*, 83; Stephen Dow Beckham in Berg, *The First Oregonians*.

6. Jette, "Kalapuya Treaty of 1855."

7. Beckham in Berg, *The First Oregonians*, 213–15.

8. Beckham in Berg, *The First Oregonians*, 213–16.

9. Salem Pioneer Cemetery, "Sanford Watson"; Copes-Gerbitz, "Defining the Historical Context of Zena Forest," 20.

Landscape and Religion in Nineteenth-Century Settlement

Vera Warren with Emily Dougan

Sitting in the kitchen of the farm house at Willamette University's Zena Farm, one has a spectacular view of Zena Forest on the gently sloping Eola Hills. Looking at the forest—or, better yet, walking along the dirt path that winds through the Douglas firs—one can begin to understand why so many Christian Euro-American settlers chose to make Zena their new home. To Euro-American settlers, Zena was an untamed wilderness waiting for human cultivation and a place for them to start a new life. To them, this place was the biblical Garden of Eden.

In the nineteenth century, as Euro-American settlers came to the Willamette Valley, they brought with them the hope of finding an Edenic promised land, and many of them found it at Zena. The historian William Robbins argues that Christian Euro-American settlers brought with them preexisting religious backgrounds and ideologies—in particular, the desire to create a new Eden. These dramatically impacted the physical and cultural landscapes of the Willamette Valley. However, Robbins fails to address the spiritual changes that the land enacted within the hearts and minds of the settlers. His analysis only describes how the land physically incentivized the settlers, and encouraged them to physically change it. His analysis does not fully engage the reciprocal nature of the relationship between man and nature. When analyzing Zena, we must take his analysis and broaden it to investigate the religious tendencies of the settlers, and how the physical landscape influenced these tendencies.[1]

In order to understand this reciprocal relationship between religion and the environment, it is important not only to study the ideals and perceptions of Christian Euro-Americans and the impacts these perceptions had on the natural landscape, but also to understand the experiences of Euro-American settlers firsthand. This chapter begins with an investigation of common perceptions of Christian settlers in the nineteenth century, and an analysis of the impact of those perceptions on the landscape at Zena. The chapter then examines the experiences of settlers in the Zena area, drawing on two bodies of material: the records of the Spring Valley Presbyterian Church at Zena (the religious center of the area) and papers from the Purvine and Walker families, members of the church and early residents of Zena. These documents demonstrate the impact the natural world had on these families. Just as religion influenced the perceptions and actions of many settlers in their interactions with the environment, the landscape, too, changed the values and behaviors of the settlers.

Religious Worldview

The nineteenth-century journey to the Willamette Valley was long and perilous. Many today are familiar with the stories of hardship and sickness that befell some of those who took it upon themselves to travel this distance. With the convenience of modern technology it is easy to take the danger of this journey lightly, but make no mistake, settlers were risking a lot making the move out West. Knowing these stories, we must ask ourselves why Euro-American settlers chose to make such a dangerous journey to Zena. Of course, many chose to make such a voyage to create better lives for themselves. Many also chose to come to the Willamette Valley with the promises of a more hospitable agricultural climate than what they had experienced back east. However, religion and religion-influenced worldviews were also powerful motivators for Christian Euro-American settlers who settled at Zena.[2]

The Christian worldview that humans have dominion over the Earth influenced many that it was their duty to settle in the Willamette Valley, or in this case, at Zena. This anthropocentric worldview likely got its genesis from the biblical story of the Garden of Eden. In this story, God creates a garden called "Eden" for his creations. He creates man, Adam, and gives him dominion over the garden and the duty of caring for the Earth and all of its inhabitants. However, man's wife, Eve, is tempted by Satan in the form of a serpent, and eats the forbidden fruit from the tree of knowledge. After discovering their sins, God banishes Adam and Eve from Eden and sentences them to toil with the Earth for eternity. In Christian scripture, this story is used as evidence of man's rightful ownership of the land, but also man's duty to labor over it as a form of penance for the sins of Adam and Eve. The story of Adam working hard as a way to repent for their sins could be why many Christians, including the Euro-American settlers from the nineteenth century, believed that working hard was analogous to praying or repenting. Thus, settlers brought with them this idea of having dominion over the Earth and needing to toil with it for repentance.[3]

For many settlers, Zena seemed like the perfect place to create this biblical Eden. As settlers began moving westward, stories of pigs that were fat and ready to eat and crops that were taller than men made their way back east. Charles Wilkes, an American explorer who came to the Willamette Valley in 1841 with the intention to develop "civilization," described the Willamette Valley as "pastoral and suitable for cultivation of a garden." These stories enticed settlers, and gave them a hope of returning to the biblical promised land. Zena, as is discussed in later chapters, offered settlers fertile soil and an excellent pastoral climate for them to create a new life. For those who settled at Zena, at first glance, these descriptions seemed a dream come true. It appeared they had found their "Eden."[4]

Shaping the Landscape of Zena

Settlers at Zena formed a largely Presbyterian community centered around the Spring Valley Church at Zena, which was founded in 1851—the year after Sanford Watson's arrival in Oregon—and attracts congregants to this day. The church building itself was constructed in 1858 with the help of several settlers discussed later in this chapter. At this church, settlers were provided with a community of like-minded individuals with whom they could practice their religion.[5]

However, the church was also used as a tool for ideological conversion. Groups within the church, such as the Home Missionary Society at Zena, worked to convert native people of the land on which Zena was founded to the ideals of Presbyterianism. The society officially started in 1887, but the group worked unofficially for many years before that. Initially, this society was intended specifically for women, and early settlers, like Mrs. Clairebourne Walker, were members. Women missionary teachers, like Mary

Gray Almira (who would later marry B. F. McLench, an early Zena settler) were also members. In addition to missionary work, the women of this society "made quilts and prepared food to sell to raise money."[6] The Home Missionary Society of Zena represented just one of many proselytizing organizations in the Willamette Valley, like the famed mission of the Reverend Jason Lee (not affiliated with Spring Valley Church), who in 1834 began missionary work to convert the Kalapuya, the main Native American group in the Willamette Valley, to Christianity. These missionaries attempted to teach the Kalapuya about Christianity while simultaneously converting their land use practices to those used by Euro-American settlers back East.[7]

With agricultural and religious conversion, settlers also physically disrupted the natural norms of the landscape at Zena. The Christian conception that man had dominion over the Earth was sometimes used to justify harmful practices. Instead of taking this idea of dominion and using it as justification to care for and respect all living things, initial settlers used it as rationale to clear out native species, depleting their populations. Settlers killed countless deer, geese, and ducks, and the provisional Oregonian government even compensated settlers for killing animals like bears, wolves, and panthers that would often hunt livestock. The settlers interpreted the Bible and other religious teachings as their God-given right to kill and use these animals for whatever purposes they saw fit. Moreover, as they believed God gave them dominion over the land to improve upon it with their agricultural practices, they saw these animals as foils to this potential, and thus saw justification to kill them.[8]

The Purvine-Walker family papers provide one such account of settlers at Zena engaging in these destructive practices. In 1845, Clairebourne Campbell Walker, his brother Wellington Bolivar Walker, and their cousin Andrew Jackson Doak and his wife Rebecca, travelled west to Oregon from Missouri. Three years later, Clairebourne and Bolivar's brother Walter Montgomery Walker traveled west to join his brothers; that wagon train also included Mary and John Purvine, into whose family Clairbourne and Bolivar would marry. Both the Walker and Purvine families originated in Illinois and were Presbyterian, and eventually would join the congregation at Spring Valley Church in 1852. In Clairebourne Walker's account, he describes his wagon trains' interactions with herds of buffalo, which Walker and other members of his party hunted during their travels. Walker describes an instance when men went out to hunt buffalo, but they could not bring back a lot of what they killed, instead leaving these buffalo to waste. In Walker's diary, it seems as though if the resources were available, they were for his family's taking, even if at times it meant being wasteful. Their journey demonstrates how even before they reached Zena, their relationship with the land was developing. Having dominion over the creatures of the land as the Bible states could have been a contributor to their wastefulness along the way to "The Promised Land."[9]

The Purvines, Walkers, and other settlers' families brought this confident approach to land use to Zena. By 1855, as Wilson Blain, a Presbyterian minister explained, Euro-American settlers had converted the "primitive environment" at Zena and in the Willamette Valley, through "extensive improvement" where "new houses, barns, and farms clothed many portions of the road." The Presbyterian ideology settlers brought with them to Zena impacted the land both physically and ideologically, through the conversion of the native people to the ideals of Presbyterianism and through the implementation of agricultural practices used back East.[10]

Shaping the People

Just as Presbyterian settlers at Zena impacted the natural environment with their religious ideologies, the natural environment influenced them. These settlers not only grew connected with the natural environment, they also established a sense of place within it. Any observer of Oregonians today can tell that a deep love and respect for the environment exists in the minds and hearts of those who live here. We must then ask ourselves how this mentality evolved. How did the sense of place that can be seen in those who live and work at Zena develop from the practices and ideologies of these early settlers? The answer is certainly not from an overt disregard for nature and a changing of the landscape with religious ideology. Thus, we must explore how the natural environment at Zena, not just existing religious ideologies, impacted Presbyterian settlers' sense of place.

The land and environmental conditions at Zena shaped the physical habits and physical religious practices of the settlers. Accounts from members of Spring Valley show their frustration during the winter months at being unable to meet regularly due to inclement weather conditions. Firsthand accounts explain that the Home Missionary Society at Zena, which was responsible for organizing community events in addition to missionary work, could not meet regularly during winter months due to "inclement weather and very muddy roads which were often impassable." By not being able to regularly congregate and worship together, the natural environment at Zena shaped and changed the settlers' sense of spiritual community. As these settlers were likely spending more time in direct confrontation with the natural environment, and becoming closer and further intertwined in it, it is likely that this disruption of group worship pushed them away from their traditional religious ideology and toward a closer relationship with the natural environment at Zena.[11]

An indication of this change can be seen through the transformation of Claireborne Walker's view of the natural world. Like many Presbyterians at the time, Walker and his family subscribed to the idea that the natural resources of the land were theirs for the taking. In his diary of their journey, Claireborne Walker expresses his annoyance that the grass, which he desired as feed for his livestock, had been "eaten off by Indian horses." This annoyance obviously represents some of the tensions between Euro-American settlers and Native Americans at the time, but it also speaks to the privilege felt by many Christian settlers at the time.[12]

The language and sentiment of that entry in Walker's diary, however, is quite different from the language in his "A Western Emmigrant's Song." Walker wrote this song in 1851 for the Spring Valley Church community as a way to commemorate and remind the congregation of the perils and power of nature and its effect on the journey of the settlers. The lyrics in this song speak of the promise of a better land, the hardships of leaving friends, deaths on the journey, and a final resignation to the land in the Willamette Valley. Walker expresses that the settlers will "quietly resign" to the forces of nature. This language indicates that Walker had come to a conclusion that he did not necessarily have the power or dominion over the land that he thought, or had been taught, he had. He realizes that the land has the power to influence him, just as he has the power to influence it.[13]

The influence of the nonhuman natural world in settler behavior is also evident in the case of B. F. McLench, who settled in the Zena area in 1850 and was a member of Spring Valley Church. Initially,

McLench resided in a small log cabin, but in 1859 he built a large house next to a large oak tree. In his study of Willamette Valley settlement, Peter Boag has argued that Euro-Americans often decided to build homes on the edges of forests because of superstitions about the dangers of open spaces—in short, they changed the land because of their beliefs. But B. F. McLench explained that he built his home at Zena near an oak tree "to take advantage of the protection afforded" by it. The natural landscape at Zena, and the strength and protective and aesthetic qualities of the oak, influenced McLench to settle by this tree, more so than his Presbyterian faith. In fact, none of the accounts from McLench or his descendants reveal religious motivation as a reason for this construction choice. In this instance, the natural landscape, not religion, motivated McLench and influenced his relationship with the land.[14]

Accounts from McLench's descendants can also give us insight into how the land impacted the religious perspective of this settler. Like many Presbyterians, McLench likely subscribed to the religious view that man had dominion over the land. However, after settling at Zena, his language in prayer reflects a different view of the land. In a speech given at the 1982 Spring Valley Church Thanksgiving Service, McLench's great granddaughter, Gertrude Hobbs, recited a prayer her great grandfather often recited at family gatherings and mealtimes after settling at Zena:

> We thank thee, now, our Father
> For all things here bestowed
> For shelter, land, and harvest
> For health and friends, and food.

In contrast to the view of having dominion over the land, this prayer seems to reflect an acknowledgment that nature is a gift. This prayer indicates that after having lived at Zena, McLench saw the value of the natural landscape, but also the potential for it not to be readily available. At this point, McLench had experienced the hardships of poor harvests and violent storms. The tree by which he had built his house even fell! These occurrences changed McLench's religious views. He began to see the land at Zena less as something to be dominated and more as an active character in the dynamic relationship between man and nature. For McLench and many other settlers at Zena, the natural environment, not just religion, created a unique sense of place at Zena.[15]

A New Dynamic

The story of Christian Euro-American settlement at Zena is unlike the normal story told about Western settlement, in which a group comes in and completely destroys and changes the natural landscape. The story of settlement at Zena is of a reciprocal relationship. The settlers did come to Zena and change the land using preexisting religious ideology as their justification. They obviously physically transformed the landscape and sought to imprint their ideologies on the people who had been living there for centuries. Effects of these changes should not be underestimated. However, the land also shaped them. The physical nature of the land disrupted their normal religious activities, and the hardships on the land challenged their notion that the land was to be dominated. Instead, settlers at Zena developed respect for the land and learned to care for it. Their sense of place became deeply rooted and intertwined with the physical landscape of Zena. This can be seen today in the beauty of Zena, and the devotion to it of the people who live there today.

Notes

1. Robbins, *Landscapes of Promise*, 92.
2. Taylor, "Climate of Multnomah County."
3. Nash, *Wilderness and the American Mind*, 8–22; Genesis 2:4–3:24 (KJV); Stoll, *Protestantism, Capitalism, and Nature in America*, 32.
4. Boag, *Environment and Experience*, 26.
5. Women's Missionary Society, "An History of the Spring Valley Church," courtesy of Eleanor Miller, Salem, Oregon.
6. Women's Missionary Society, "An History of the Spring Valley Church," 8.
7. Women's Missionary Society, "An History of the Spring Valley Church," 10; Loewenberg, *Equality on the Oregon Frontier*, 35–60; Brosnan, *Jason Lee, Prophet of the New Oregon*, 71.
8. Boag, *Environment and Experience*, 10–16.
9. "Purvine-Walker Papers," 4–19, courtesy of Eleanor Miller, Salem, Oregon.
10. Boag, *Environment and Experience*, 146.
11. Women's Missionary Society, "An History of the Spring Valley Church," 8.
12. "Purvine-Walker Papers," 32.
13. Women's Missionary Society, "An History of the Spring Valley Church," 23.
14. Hobbs, "Thanksgiving Service Spring Valley Church Speech," courtesy of Eleanor Miller, Salem, Oregon.
15. Hobbs, "Thanksgiving Service Spring Valley Church Speech."

From Wilderness to an Agricultural Landscape

Amanda McClelland with Andrew Spittler

For nineteenth-century Euro-American settlers, the Willamette Valley promised the potential for a new Eden, but migration and settlement also entailed substantial risk to both life and property. In 1850, William S. Pickrell wrote to his wife's brother Sanford Watson, who had recently left Illinois in search for a piece of prime Willamette Valley farmland:

> Dear Brother:
> It is now near one year since you left Springfield, and we have not had a line from you. We are now anxiously looking for letters, as several have received them from friends who started with you last Spring. . . .
>
> The gold mania has not subsided, it is thought by some that there will be a larger emigration to California this year, not many, however, from Springfield. I do not hear of many going to Oregon, though from the last accounts we get from there I should think the latter place the most certain for a fortune. A great many who went to California have returned and from what I can learn, two-thirds have come back poorer than they went. A great many have died.

Sanford Watson would find his place in the Eola Hills of the Willamette Valley, on a 640-acre claim that eventually became part of Zena. So would his neighbors Thomas Warriner and Edward Robson, the other initial Euro-American residents of Zena. These men, their families, and their community endured many hardships in their effort to transform the Willamette Valley into a productive and profitable agricultural landscape. Drawing on Peter Boag's analysis of Willamette Valley settlement and community, this chapter will explore how Watson, Warriner, and Robson sought to transform their land, their place and reputation among the larger Spring Valley community, and how they saw themselves in the surrounding landscape. Through this process, the Watsons, Warriners, Robsons, and their neighbors would change not only the lands in which they lived and worked, but also themselves, their connection to Zena, and their sense of place.[1]

Migration and Motivation

Thousands of families, including the Watsons, Warriners and Robsons, traveled west to the Willamette Valley in the mid-nineteenth century. In 1849, Sanford Watson sold his family farm in Springfield, Illinois, and began the long journey to Oregon with his wife, Maria, and their three children, Virginia, age 10; Henry, age 8; and Sanford Jr., age 5. Maria's brother, Alfred Elder, traveled with them as well. From the spring of 1849 to the dead of winter, the Watson family traveled by wagon train to Polk County, Oregon. Sometime, either in the midst of travel or very soon upon arrival, Maria gave birth to their fourth child, Cecilia.[2]

Like Watson, Thomas Warriner had traveled from Illinois to Zena. He had been granted 160 acres of land after serving under Captain Maxon's company in the Oregon Militia during the Cayuse War. Thomas Warriner and many other settler-soldiers like him literally fought for their land, yet received it in the form of claims hundreds of miles away, far from the Whitman Mission and the land formerly inhabited by the Cayuse. When Warriner was done serving, he brought his daughter Melinda and his brother W. C. Warriner by covered wagon to Zena. He spent the rest of his life farming his land in Oregon.[3]

Edward Robson first made his way to America by sea, voyaging from Worcestershire, England to the San Francisco Bay in the early part of the century. From there he traveled north through the dry heat of the Sacramento Valley towards the Oregon Territory (most likely in spring). He finally entered Oregon's lush valley of the Willamette in September of 1844. Being an immigrant from England, he had challenges getting a Donation Land Claim in Oregon. Ten years would pass before he received United States citizenship and legally obtained his few-hundred-acre claim at Zena in 1855.[4]

Difficulties from the Outset

For emigrants and immigrants alike, the journey to Oregon was a long and difficult one. Some didn't make it, and if they did, much of their property would have been lost while passing through the Cascade Mountain Range. Many travelers underestimated the inevitable challenges that came with crossing the Cascades. As a result, most arrived in the Willamette Valley later than anticipated, hungry and without many of the necessities for beginning a homestead. Hunting prospects in the valley were meek for newcomers, as the available game had rapidly declined after the first Euro-Americans moved in. For the first several months, settlers were made to survive on boiled wheat, potatoes, and salted pork.[5]

Possessing little food, no shelter, and only a fraction of the livestock they left home with, many settlers soon acquired a bitter relationship to the land that reflected a fierce contempt for the rough conditions of a cruel wilderness. This mindset pitted many against their environment, rather than putting them in a position of care, appreciation, and stewardship for the valley's natural resources. Incredibly, many survived the harsh conditions of the frontier and were able to continue in the American dream of westward expansion. However, memories of those first cruel months fostered generations of animosity against the destructive might of the landscape.

Fortunate location of a claim upon the landscape was crucial for the very survival of early settlers. According to an 1852 Spring Valley surveyors' map, the claims of Sanford Watson, Thomas Warriner, and Edward Robson were established near features of the Zena landscape essential to the initial survival of themselves and their loved ones, as well as their future prosperity upon the land. For example, the Watson's claim was mostly comprised of open prairie, but the eastern side consisted of forested hills. Sanford and his young family had finally found the prime farmland they'd been seeking, and it would soon serve as a source of income. However, without shelter built from timber found in the eastern hills, Watson's family would have had difficulty surviving their first Oregon winter.

When the Watsons arrived, it would have taken them a few months to erect a one-room cabin built from fir logs and a cedar shake roof. The forested land at Zena teemed with natural building resources, enabling the Watsons to construct their home in much the same way they would have back east. When the home was completed, the settlers' attention immediately shifted to agricultural improvements. Planting a family garden was first on the list. The garden would provide food crucial to the family's survival over the next year. The fertile valley supported gardens of abundance, planted with potatoes, cabbages, peas, turnips, onions, parsnips, tomatoes, carrots, corn, and beans. Settlers also grew chicory as a coffee substitute, as well as their own tobacco.[6]

However, a thriving family garden was not what the Watsons and others like them had traveled 2,000 miles to get. Instead, Watson, Warriner, and Robson were in part attracted to Zena's market geography. Farmers could easily sell their products to the growing towns of Lincoln and Salem, and they could also export their goods, thanks to Zena's relative close proximity to the Willamette River. This prompted farmers such as Sanford Watson to begin cultivating as much of their land as possible as soon as they could. However, it was no easy task. Breaking the grassland was done with a crude, ox-drawn plow. The plow would break often, and the one or two surviving oxen would fatigue easily, still weak from the long journey. This made for slow going. There was a scarcity of cattle and horses in the valley, due to the enormous obstacle of the Cascade Mountains. So, the settlers had to make do with what they had. During the plowing season, the men would have cursed the soil that stayed wet and muddy long into spring. Plows broke, and oxen sank deep into the sopping ground. Although the soil was fertile from the flood sediments, these same floods could turn the rich soil into marshlands and mud. Much of the soil did not naturally have good drainage. The valley's enormous amount of winter rain angered and challenged the farmers, many of whom were used to the much drier winters of the Midwest. However, they battled these challenges, and once the ground was plowed, all of the seed was sowed by hand. This took many hours, and was quite inefficient. The seed would have been spread unequally and in a way not optimum for the wheat's growth. In the first year, seed was particularly hard to come by, as most would be planting wheat, and it would have been a struggle in those winter months not to eat the seed they were supposed to be saving. If they failed to save seed for planting, some new settlers would work on established farms in exchange for wheat seed. This common exchange would make wheat the legal tender of Oregon in the early years of settlement.[7]

When the seeds were in the ground, the settlers made farming implements, which included improving plows, sharpening tools, and building carts and other attachments for teams of oxen or horses. They also built facilities such as pens, troughs, or threshing floors. As time went on, barns would have been raised to protect livestock, and sheds would have been built to store tools and other farming equipment. Again, all of these building resources were abundant at Zena, and the settlers took freely what they needed from the forests. At Zena, Watson, Warriner, and Robson might have helped each other with barn raising and other large projects. These three men and their families worked together to continue to build the growing towns in the area. For example, Sanford Watson opened up a US post office in Spring Valley in 1852, helped petition for (and became a trustee of) Bethel College, and served as the director of the Bethel School District from 1862 to 1865.[8]

Even though the three men were part of a growing Spring Valley community, many hours of labor were spent in solitude, the only company being the vast, unfamiliar landscape surrounding them. This one-on-one interaction with the land must have had a profound effect on their sense of place. Living within neighboring distance of one another, the three men faced similar challenges when interacting with the landscape at Zena, such as immense flooding of grasslands and the crude farming implements of the time. They would have had to learn first to adapt to the landscape's unfamiliar features and cycles in order to reap the benefits of their meticulous labor. Long periods of solitude upon the land may have served as a time of learning and reflection. Frustration with their own physical limitations must have tormented these men, as each of them had brought loved ones whose livelihoods depended on their success. In this way, Watson, Warriner, and Robson collectively shared a sense of place that viewed their land as not only a means of economic prosperity, but a means by which to care for those they cared about: a wife, a child, a sibling, a parent, or even a friend.

However, these men were also individuals, each driven to Zena under different circumstances and with different family and community connections. In Sanford Watson's case, after ten years of marriage and the birth of three children, he decided to sell the family farm in Illinois and start over, perhaps with the hope that Oregon presented greater opportunities to his kin. Soon after his arrival he opened up a business, for as a farmer in Illinois he knew it would be some time before the fertile land would be prosperous, and he would need an alternate source of income. However, in deciding to open a US post office, Watson was not only providing for his immediate family, but also providing the wider Spring Valley community with a convenient connection to the rest of the country, enabling better business, education, and familiar correspondence.

Watson not only sensed Zena as a place in which to grow his business and family, but also as a community—a place in which to connect, share, and prosper with those surrounding his young family. In contrast, Edward Robson made his way to Oregon after being estranged from his family in England. Now half a world away and without a wife or children of his own, Robson must have had an alternate sense of Zena, one of individual opportunity as opposed to the familial opportunity sensed by men like Watson or Warriner.[9]

Livestock: Food and Profit on the Hoof

Over the course of a few laborious years, farms became more established and production began to grow. Along with vegetables and grains, Sanford Watson also raised pigs, cattle, and horses. Pigs were an important part of settlers' farms. Initially, pigs were able to survive off of acorns and camas root. It was essential in the beginning that the farmers didn't have to use their own resources to feed the pigs. However, in the long run, pigs also contributed to increasing the efficiency of the crude farm practices. Farmers were able to feed the pigs the skim milk and whey from the cows that usually would have been discarded. The pigs were also able to utilize the grain, potatoes and other crop remnants left in the fields due to inefficient harvesting methods. In essence, pigs on the farm helped to create a less wasteful food system in which the pigs could consume what would usually be waste, and then the people were able to eat the pigs. However, in the 1840s and early 1850s, animal slaughter and meat preservation was a challenge to farmers because many lacked the necessary implements, and salt for preserving was hard to come by. As trade and population in the area increased, raising animals for meat became more practical.[10]

Cattle were also raised at Zena, and the sprawling grasslands were of enormous benefit to the herds of cattle. One of the main reasons the land was so good for grazing was because of Native Americans' prescribed burnings of the oak savanna. However, when the Euro-Americans settled the area, they neglected to continue the burning practices. As a result, many places where the grass had once flourished gave way to trees and shrubs over time. But while it lasted, the Willamette Valley was deemed one of the best places to raise cattle. The mild climate kept the grass green all year, and allowed farmers to have cattle without worrying about over-wintering them in shelter or having a supply of grain to feed them during winter months. In the early years of his farm at Zena, Sanford Watson would most likely have had a Spanish breed of cattle, which were hardy animals—able to survive without much shelter—and produced excellent beef. Their only drawback was that they produced little milk, and a good dairy operation would be impossible with the Spanish breed. Over time, huge herds of many other breeds were trailed northward from California into Oregon. With these different types of cattle, dairy herds were soon established in Oregon, and milk, cheese, and butter production became an important part of the local economy.[11]

From Soil to Market

Of all crops grown in the rolling hills of Zena, perhaps the most important was wheat. Wheat was the area's chief export, and for frontier farmers, it was what paid for the few necessities and small luxuries that were important for survival. The two main varieties of wheat that were cultivated were spring red wheat and white seed wheat. Growing wheat in such a wet climate was a challenge. Farmers dug ditches around the acreage planted in wheat in an effort to drain the fields of their winter water. The average wheat harvest was around twenty bushels per acre. However, effective harvesting of all the wheat that was grown was difficult. All harvesting was done by hand, and two men such as Sanford Watson and his brother-in-law Alfred Elder would have been able to cut and bind approximately three to four acres of wheat per day. In 1848 the first mechanical thresher was brought to the valley, but it took a while before they were widely available to farmers throughout Oregon.[12]

Although not a particularly popular crop among farmers at Zena, orchards of all kinds nevertheless played an important role in the valley's agricultural production. By the 1860s many of the more improved farms had young orchards. The common fruit trees that were grown included apple, pear, cherry, and hazelnut trees. Even today there are remnants of the early Oregon orchard industry at Zena. Northwest of the farmhouse is a hillside dotted in fruit trees, mostly apple. Scattered around the farmhouse are a few trees, separate from the hillside. These may have been part of a "kitchen garden" much like the Watson's garden in their early years of settlement. Early on, the lack of canneries forced the growers to sell fruit fresh to local markets, which was a hindrance to how much the orchard economy of the valley could grow. Soon, though, transportation technology improved tremendously with the capacity to keep produce refrigerated on trains, and that allowed fruit producers to gain markets all over the country.[13]

New transportation and farm technologies allowed and encouraged Willamette Valley farmers to increase agricultural production. In the 1850s, the primary mode of transporting large amounts of goods was by boat, either on rivers or the ocean. For Sanford Watson, living at Zena in Spring Valley west of Salem, he would have had to load up a large wagon of produce and make the journey into town with a team of horses. In Salem, his produce might have been bought and sold, and perhaps put on a barge traveling north to Portland.

But major changes in the American agricultural sector were underway, as Sanford learned from his brother-in-law:

> Illinois is more prosperous than perhaps at any former period; the influence of the operation of the railroad has given a different aspect to things in the surrounding country, in the way of lumber especially.
>
> They slaughtered a great many hogs here last Fall, the stock on the road from Springfield to Alton has been taken and the work commenced. There seems to be considerable excitement in N.Y. on the subject of railroads.[14]

By the 1870s, telegraph lines and the Oregon Central Railroad had connected Portland and Sacramento, and in effect the entire country. The railroad dramatically increased how much could be exported and imported to the Willamette Valley. In 1868 the first shipment of several hundred barrels of flour was sent to New York, which was part of the efforts of Portland traders to establish distant marketing outlets for the increasing wheat surpluses of the Willamette Valley. Some of the wheat produced by Sanford Watson or Thomas Warriner might have been included in that shipment, and for them, that meant they were able to expand the amount of land they had under wheat production, and still have the ability to sell it all. This connection of the United States by train truly marked the shift between local economies and global economies. The development and expansion of the region's market economy initially lessened the farmer's burden of over-producing crops, and would eventually allow many to expand their agricultural production.[15]

Conclusion

The Zena community grew in both population and agricultural business, and new advances in transportation and agricultural technology would encourage farmers to further assert dominance over the nonhuman natural world. But in the nineteenth century, farmers like Watson, Warriner, and Robson had no choice but to learn how to cope with Zena's environment: its heavy winter rains, followed by flooding, followed by the muddiest of roads to trading towns. In death these pioneer-farmers left behind a legacy of growth and agricultural development in Zena's Spring Valley community that, out of necessity, worked with the landscape. And despite the changes brought by the twentieth and twenty-first centuries, the essential lesson of the nineteenth century—that humans must in many ways abide the nonhuman natural world—would remain the same.

Notes

1. Pickrell, Letter to Sanford Watson, 61–63; Oregon Provisional and Territorial Government, Land Claim Records; Boag, *Environment and Experience*.

2. Salem Pioneer Cemetery, "Sanford Watson"; Oregon Provisional and Territorial Government, Land Claim Records.

3. Oregon Provisional and Territorial Government, Land Claim Records; Copes-Gerbitz, "Defining the Historical Context of Zena Forest," 20.

4. Oregon Provisional and Territorial Government, Land Claim Records.

5. Gibson, *Farming the Frontier*, 135.

6. Copes-Gerbitz, "Defining the Historical Context of Zena Forest," 18; Bunting, "The Environment and Settler Society," 418.

7. Bowen, *The Willamette Valley: Migration and Settlement*, 74, 79; Bunting, "The Environment and Settler Society," 416.

8. Gibson, *Farming the Frontier*, 135; Salem Pioneer Cemetery, "Sanford Watson."

9. Salem Pioneer Cemetery, "Sanford Watson"; Oregon Provisional and Territorial Government, Land Claim Records.

10. Copes-Gerbitz, "Defining the Historical Context of Zena Forest," 18; Bowen, *The Willamette Valley: Migration and Settlement*, 87.

11. Bowen, *The Willamette Valley: Migration and Settlement*, 60, 83; Gibson, *Farming the Frontier*, 129.

12. Gibson, *Farming the Frontier*, 142; Federal Cooperative Extension Service, "Oregon's First Century of Farming."

13. Wright, "An Economic and Biographical History of Heirloom Apple Trees at Zena Farm," 24.

14. Pickrell, Letter to Sanford Watson.

15. Boag, *Environment and Experience*, 136–39.

Transformative Agricultural Technologies at Zena

Emily Schlieman with Keller Cyra

Prior to the revolutionary technological breakthroughs that distinguished the twentieth century from the nineteenth century, the nonhuman natural world held the upper hand in its relationship with people. Farmers used steel tools powered by draft animals to plant their crops, thereby exercising a limited amount of control over the land, but still felt that they were ultimately subject to the powers of the natural world. As scientific knowledge expanded and paved the way for the creation of more powerful technology that allowed people to have a greater impact on the landscape, they felt their ability to control their natural surroundings expand. New technologies, from drainage systems to machinery, allowed farmers to significantly expand the size of their farms and the variety of their produce, and the high expectations of continual progress in the field of science left them with hope for more powerful domination over nature in the future.

At Zena, clay tile drainage, which was installed in the 1960s, demonstrates human efforts to overcome nature. Tile drainage is a method of subsurface drainage that alters the composition of the soil by reducing its water content. This type of technology has been applied to the Zena landscape, as it has elsewhere in the Willamette Valley, in areas deemed too saturated for easy cultivation. Soil drainage promised to free people from some restrictions set by the natural world, empowering them to choose where and what they wanted to farm, instead of having nature determine those aspects for them. William Robbins, in his environmental history of Oregon, *Landscapes of Conflict*, demonstrates how other technologies, especially pesticides and particularly DDT, held out similar promises of mastery, only to result in unexpected and uncontrollable consequences. The story of tile drainage at Zena reveals a similar story, showing the leaps that humans were making in their efforts to free themselves from the constraints of the nonhuman natural world—

but also the ways in which the nonhuman natural world reacted to those efforts.[1]

Advances in technology made in the twentieth century caused a radical shift in sense of place, apparent at Zena, as people became more confident in their ability to master nature and realize the "promises" of abundance in the West. Of course, the first Euro-American settlers also had the mindset to reap as much from the earth as possible. What technology added to this dynamic was a fuller capability to do so, allowing people to become more exploitative of their environment. Innovations such as tile drainage, pesticides, and machinery, significantly altered the relationship people had with the nonhuman natural world. But at the same time, those technologies reaffirmed a previously formed sense of place that people could potentially conquer nature. People entered the twentieth century with the determination to overcome natural dilemmas that were at one time insurmountable given their primitive technology, but believing that with technological advancements they could defeat these natural boundaries. Farmers in the Willamette Valley saw science and technology as the means to achieve a more favorable relationship with their environment, a relationship in which farmers were not restricted by forces of the natural world.

Tile Drainage

In their efforts to master agricultural hydrology, Willamette Valley farmers found an enthusiastic ally in the state land-grant university, Oregon College of Agriculture (now Oregon State University). Its Experiment Station made many contributions to local farmers and played a large role in what path state agriculture would take in the mid-1900s. The College of Agriculture provided farmers with research and suggestions for innovations specific to the Willamette Valley, which was helpful for local farmers. Scholars at the College investigated methods of managing weeds, pests, predators, and plant diseases, which were farmers' chief combatants. And they advocated agricultural drainage, which, as a 1916 Oregon Agricultural College Experiment Station publication estimated, could help improve productivity on nearly two million acres of "wet land" in the Willamette Valley and other valleys in western Oregon. Through experiments, articles, and assistance to farmers in installing their drainage systems, the Experiment Station helped bring the latest technology to the waterlogged fields of the Willamette Valley.[2]

9. Map of tile drainage at Zena; the diagonal lines on the left side of the photo represent the location of the tiles. *Courtesy of National Resource Conservation Service, Dallas, Oregon*

The Experiment Station's 1916 publication on wet lands explained tile drainage as a useful innovation in which "excess water finds its way to the tile drains through the line of least resistance in the soil and enters between the ends or 'joints.' This process of drainage does not leave the soil without moisture but simply removes the excess and makes room for more usable moisture and some air." The idea remains essentially the same seventy years later.

In short, farmers install a series of short pipes—"tiles"—in their fields to decrease the water content of the soil and encourage crop growth. Tiles have been composed of different materials over time: the earliest were made of clay—like those at Zena—but cement, wood, or metal began to replace clay in the late 1960s and '70s. The individual tiles are quite short, typically a few feet long, and lay stacked with space in between each other so that the water can get into the pipe. When the water accumulates into a concentrated flow, it flows to an outlet, which is typically a stream to which the tile system connects. These tiles are usually set four feet down, and become shallower as it nears the stream serving as its outlet. The spacing of tiles depends on the type of soil it drains. Tiles in clay soil would be much closer together than those in sandier soils since water would percolate through the denser clay and reach the tiles more slowly than the sand. According to the USDA, the soil at Zena is hazelair silt loam, which is "moderately well drained," while the soil to either side in the hills consists of slightly different variations of silt loam, a type of soil not too high in clay but labeled instead as "well drained."[3]

Advances in technology have made tile installation much easier and efficient than past methods. According to local farmer Roger Loop, the method of placing tile drainage earlier in the century disturbed the soil much more than the refined machinery of the latter part of the century. Originally, workers dug trenches with a backhoe to place the tile in the ground, and filled it back in afterward, which was took a lot of time and money. Later on, tools called subsoilers were able make smaller slices in the ground and were not only less intrusive, but also more economical and less time-consuming. Reels were then used to lower the tiles piece by piece. These innovations made the process faster at a higher cost, but this did not steer farmers away from this solution to more crop-friendly soil.

Tile drainage was an appealing alternative method to topical ditch drainage when modern machinery first entered the scene in Oregon in the 1930s. Moving drainage beneath the surface allowed the large and heavy farm machinery to operate in the fields without fear of disturbing the drainage channels. Farmers also preferred subsurface drains because they required little maintenance and they did not detract any land from production. There were 320,000 meters of subsurface drains installed in Oregon between 1872 and 1916. Researchers in the Experiment Station observed that drainage is "one of the most permanent improvements that can be put on the land," which appealed to farmers because it raised the land's market value.[4]

Farmers who installed tile drainage in their fields generally reported satisfaction with the outcome, saying it boosted their production levels and therefore their income. Although tile drainage is expensive to install—it was 300 to 400 dollars per acre in the 1960, but now is closer to 1,000 dollars per acre—it pays off quickly, as farmers attest in their praises. In the 1914 Experiment Station's publication about drainage, there is over a page of farmers' praises regarding the benefits of tile drainage. One farmer glowed, "I consider that tile has doubled the output of my land and the system is not yet complete. I certainly do consider that it pays." Another noted the ecological difference it made and how it helped his crops: "After the soil is thoroughly drained, it becomes very porous and will hold the moisture in dry weather fully one hundred per cent better. The soil becomes warm and the rootlets of crops can penetrate the soil a great deal deeper." These farmers are particularly emphatic about how tile drainage contributed to their agricultural efficiency.[5]

Good drainage also prolongs the growing season and allows farmers to grow different crops than would have originally been a match for the wet soil. Loop pointed out that although Oregon is a good place to cultivate, nights are cool during the growing season, and so farmers need the longest possible growing season. Tile drainage provides this extra time. Additional benefits of tile drainage include making it easier to use heavy modern machinery because it reduces the amount of water in the soil, which is especially useful in the case of early rains at the end of harvest season. The Experiment Station also put forward that drainage would add 30 million dollars to annual agriculture production as well as increase land value in the state, giving the state economic incentive to promote drainage.[6]

Since the tiles at Zena are made of clay, they were most likely installed before the late 1960s, when plastic became the most popular tiling material. An undated map of tile drainage at Zena does not show topography, but the tiles are located in the natural valley between two hills to the southwest of the current farm, and cover a length of approximately 2000 feet. Since the tiles run parallel to the natural drain, or valley, they utilize and benefit from the natural lay of the land. Technology and the natural environment are thus working together.

It appears that the Zena design is pattern tiling, which Loop described as a series of smaller lateral pipes that feed into a main pipe that then flushes all the accumulated water to the outlet. Another tiling design that Loop mentioned is a one-line system, in which the tiling is at the bottom of a draw, or slope. This design in particular helps with erosion problems, Loop noted, and has helped reduce the amount of erosion occurring in the rolling Eola Hills.

Today, the drained land at Zena is not being utilized in the same way as when the tiling was installed. Instead of growing crops, such as wheat or fruit trees, Douglas firs now grow there. This change is likely a result of Zena's change in hands over the years. The shift to commercial forestry was probably made by the Zena Hills Timber Company that bought the property from the Higley family in 1990, and from whom Willamette bought Zena in 2008. Zena's landscape has moved increasingly toward forestry and away from agriculture, as the surface of the tile drainage demonstrates. The tiling does not serve its intended purpose any longer, and does not appear to perform an adequate job with the current plant growth on its surface.[7]

In a visit to Zena the authors made in October, we located the site of the clay tiles and observed the plants growing on top of the drained soil. Some colleagues noted that the Douglas firs in the valley were not as large as the Douglas firs of the same age on higher ground. According to this observation, it appears the tiling was not bringing the water level in the soil down to an ideal for maximal Douglas fir growth. This demonstrates the opposing dynamic of these two actors—humans with technology, and nature—and each is determined to have their own different results. Humans have used technology to alter their landscape to become more beneficial according to their values, in this case, productivity and efficiency. Tile drainage at Zena is a clear illustration of the reciprocal push-and-pull relationship between humans and the nonhuman natural world.

Inevitably, technologies have imperfections and limitations, and cannot always counteract the natural forces they fight. Pipes can burst due to pressure from blockage and too much water, and since they are subsurface structures, they are more difficult to access and fix. If this happens, as Zena area resident Anne Walton points out on her property, mounds form on the surface where the pipe underneath burst. However, it is one of the missions of science, or the expectation many hold for this field of study, to pursue the perfection of technology, so improvements are constantly in progress.

Tile drainage was clearly a popular method of transforming marshy land into an efficient agricultural landscape as early as 1914 in Oregon, and has remained a useful method of field drainage through the latter part of the century. Sue Reams noted that farmers tile their fields despite having well-drained soils, both to extend their growing season and because the ability to control the amount of water in the soil allows farmers to have more control over the crops they can grow in specific areas. It seems that manipulation of the natural world has magnified as demand for more efficient farms has escalated over the years.

There is a constant interplay between humans and nature in which humans utilize technology in their efforts to overcome natural limitations. People's sense of place has changed particularly because technology has changed their place. It has expanded possibilities for many, creating options for altering their environment and garnering the maximal amount of produce the land has to offer with some help. In addition to transforming the physical place people inhabit, it dramatically changes people's relationship with their environment, allowing humans to make advances in overcoming some restrictions set by the nonhuman natural world, though it is never clear who wins in the end.

Farming Machines

Farm machinery provides another example of how technology made a dramatic impression on the land and on those who sought to master it. Zena neighbor Anne Walton has a photo from 1911 of her grandmother's friend riding a tractor of the day. This made planting the grass seed and grains that Walton said grew well in the Zena area much easier and faster. And those improvements in productivity and efficiency continued throughout Oregon, as evidenced by advertisements in the *Capital Press*, a weekly agriculture-focused newspaper published in Salem since 1928. It is interesting to trace the evolution of farm technology through the advertisements in the paper over the decades, especially when there was more advanced machinery following the 1940s. These advertisements for tools such as a Ford tractor appealed to farmers' sense of

10. Farming in the Zena area, 1911. *Courtesy of Anne Gilbert Walton, current resident of Zena Springs Farm, Zena, Oregon*

efficiency in time and work. The advertisements contained blurbs such as "Get more work done in less time! Economical. Low Cost. Heavy Duty. Powerful," for Scotty Riding Tractors, and in a Howard Rotavater advertisement, "Tills an Acre in Less Than 2 Hours!" The number of machinery advertisements in the *Capital Press* signifies the general popularity of modern machinery in the mid-1900s forward. According to an article in *Power to Produce*, the 1960 Yearbook of Agriculture, "An average of 7 acres could be planted in a 10-hour day with the one-row, one-horse planter when the rows were 3.5 feet apart. With the same width between rows, a six-row tractor planter can cover 80 to 100 acres in a day," demonstrating the leap in efficiency between the two generations of machines.[8]

Such efficiencies transformed the appearance of agricultural landscapes. While the number of farms in Oregon decreased, from a peak of 65,000 in 1935 to 54,000 in 1958, farms increased in size because modern machines could cover more ground in less time. In correspondence, the rural farming population decreased around this time, because there were fewer farms, and engine-powered machinery replaced manual labor of humans. Zena participated in the agricultural consolidation phenomenon spurred by modern technology in the beginning of the twentieth century, when Daniel J. Fry purchased Thomas Warriner's land upon his death in 1889, and then added Sanford Watson's and Edward Robson's land in 1904 and 1909. Fry's actions were significant not only because he created the plot of land that is called Zena today, but because he contributed to the trend occurring in other parts of Oregon and the United States of farm expansion.[9]

Conclusion

Like other places in the Willamette Valley and beyond, Zena has been touched by dramatic developments in agricultural technology. Technology has revealed possibilities that were not realized before, such as making better agriculture soil in a marshy area, which raised confidence in humans' capability of controlling their environment in other ways. However, nature remains fickle, and it does not always comply with people's technological forces, as shown by the stunted Douglas firs. The result is a hybrid landscape of technology and nature. As environmental historian William Cronon puts it, "Nature is not nearly so natural as it seems. Instead, it is profoundly a human construction." At Zena, that "human construction" took place both above ground, through farm machinery, and below ground, in drainage tiles, creating a landscape that both reinforced and challenged a sense of place rooted in efforts to master the nonhuman natural world.[10]

Notes

1. Robbins, *Landscapes of Conflict*, 114.

2. Powers and Teeter, *The Drainage of "White Land,"* 10.

3. Robbins, *Landscapes of Conflict*, 16; United States Soil Conservation Service, *Willamette Valley Drainage Guide*, 52; USDANRCS, "Web Soil Survey"; Sue Reams (USDANRCS, Dallas, Oregon), in discussion with Roger Loop, 2012. Reams adds that any fertilizers and/or pesticides that farmers apply to their crops can drain into the tiles and accumulate in the streams. This may contribute to surface and ground water quality issues in the Willamette Valley and other locations where tile drainage is installed. Another danger she suggests is the inadvertent draining of wetlands.

4. Backlund et al., "Effect of Agricultural Drainage on Water Quality," 289; United States Soil Conservation Service, *Willamette Valley Drainage Guide*, 12; Powers et al., *The Drainage of "White Land,"* 16.

5. Powers et al., *The Drainage of "White Land,"* 16.

6. Backlund et al., "Effect of Agricultural Drainage on Water Quality," 290; Powers et al., *The Drainage of "White Land,"* 10.

7. Copes-Gerbitz, "Defining the Historical Context of Zena Forest," 20.

8. Ad pages in *Capital Press*, 4 May 1956; Hudspeth et al., "Planting and Fertilizing," 147.

9. Federal Cooperative Extension Service, *Oregon's First Century of Farming*, 2; Copes-Gerbitz, "Defining the Historical Context of Zena Forest," 20.

10. Cronon, introduction to Robbins, *Landscapes of Conflict*.

Evolution of Land Use Planning in Oregon

Kyle Carboni with Morgan Gratz-Weiser

Throughout the beginning of the twentieth century, and especially after World War II, the Willamette Valley underwent substantial population growth. Initially this growth was unregulated, and there was mounting public concern that encroaching urban development would cause environmental degradation and loss of valuable agricultural land. In response, the Oregon state legislature initiated land use planning guidelines, a process that spanned decades and was rife with tension due to disagreements between urban environmentalists and rural farmers. Much of the debate occurred based on the details of implementation, as most were in favor of some sort of land use planning. However, the urban proponents of conservation held differing ideals than that of the rural land users.

This conflict in land use legislation materialized in the late 1960s and the early 1970s when the fear of urban growth and commercialization prompted the creation of Senate Bill 10 in 1969 and Senate Bill 100 in 1973, which mandated statewide land use planning. The differing perspectives in land use arguments can in large part be attributed to social and economic bias, based on location, and the duration of sense of place with the land. Many urban residents viewed undeveloped areas as a place to escape from the crowded cities in which they lived and worked. There was much urban support for land regulation and zoning with the desire to protect city aquifers and maintain the natural majesty of Oregon by preventing unregulated urban expansion. The opposition to land use laws came from small farmers, lumbermen, and working class people often located in sparsely populated rural areas. They depended on the land for their livelihood and did not want it be regulated by the government, which seemed more concerned about the fears of environmentalists and the desires of commercial industries than local citizens' welfare. These conflicting ideologies formed a gap between the groups, further exaggerating the tension of the land use discussion. It is often forgotten that most parties involved agreed on the same tenets of land use planning, yet disagreed on the methods of implementation.

A number of scholars have investigated conflict over land use planning. In *Oregon Plans*, Sy Adler lays out the path to land use planning from a 1947 statute that allowed Oregon counties to adopt and create land use plans, to the drafting of Senate Bill 100 and its implementation. Adler shows how land use planning was complicated and filled with contention between political activists, city planners, elected and state officials, and rural communities. H. Jeffrey Leonard takes a similar perspective in *Managing Oregon's Growth: The Politics of Development Planning*, dissecting the debates between state and local governments regarding the implementation of land use planning laws, and efforts to revise or even rescind the regulations of 1973. Adler and Leonard's analysis in many ways confirms the findings of Jerry Medler and Alvin Mushkatel, who in 1979 conducted a study on the relationship between the voting habits of urban and rural communities in relation to Senate Bill 100. They concluded that the cause of disparity between urban and rural communities was due to the socio-economic status of the two populations in that area. Those in favor of land use policy were generally from higher socio-economic levels, and were from areas that experienced rapid growth during the previous decades, while opposition stemmed from lower income levels in areas of less drastic development.[1]

Adler, Leonard, Medler, and Mushkatel's arguments are largely borne out in this chapter's review of land use planning in Polk County and in the Zena area more specifically. The Zena area is comprised of agricultural land split between old family farms and newer hobby farms and vineyards. Whether the land is used more commercially or on a small scale, the implementation of land use planning has a variety of implications. Using personal interviews with Zena area residents, as well as analysis of Polk County's first comprehensive land use plan, this chapter brings a much larger debate closer to home, as we can see how such regulations have altered and limited land use in the Willamette Valley.

Zena and Polk County, although only a small stage in the state—and in some cases, nationwide—debate about land use planning in the twentieth century, serves as a microcosm of a deep-rooted conflict. As land use planning has been implemented in the Polk County area, people from different backgrounds and with different interests have reacted in different ways, sometimes generating conflict. This conflict often stems from their differing senses of place—their way of relating to and connecting with the Zena area.

The Beginnings of Land Use Planning in Oregon

Land use regulations stretch back to the early twentieth century, although a comprehensive system did not develop until the middle of the century. In 1919 and 1923 Oregon developed legislation that granted cities the authority to develop land use regulations. Between 1930 and 1940 the Willamette Valley in Oregon experienced a population increase of nearly 40 percent, thanks to migration during the Great Depression and World War II. Land use planning remained restricted to cities until 1947, at which point urban expansion had become so extensive that the legislature was prompted to extend planning authority to the counties. The Oregon legislature created a statute in 1947 that allowed counties to create land use plans and ensure that proposed zoning regulations were in accordance with the plans. Unlike cities, counties were required to form planning commissions, which evaluated local land and provided recommendations for development patterns, also known as comprehensive plans. Over the next decade there was little progress, with only six counties adopting land use plans. However, in nearly half of the counties, there were areas subject to planning discussions based on requests and petitions by local land owners. By the 1950s there was growing advocacy for the zoning of farmland, as the Willamette Valley population was soaring and an estimated four acres of farmland was being lost for each acre of urban development. By the 1960s, zoning and planning were usually voluntarily in implementation and in some areas were merely advisory. Land use planning was essentially for the purpose of maintaining established land use patterns and protecting middle- and upper-status neighborhoods from land-use changes. In 1961 the state legislature authorized the lowering of tax assessments for land that was zoned for farm use, in an attempt to maintain the land for agriculture.[2]

Although only a small stage within the larger picture of land use conflicts in the later decades, Zena serves as one example of the contention regarding land use planning. In the 1960s Polk

County experienced a rapid population growth of 33 percent. This growth was concentrated within urban centers including the West Salem, Spring Valley, and Rickreall areas. West Salem, one of the closest urban centers to Zena, posed a legitimate danger through uncoordinated rapid expansion. In 1954 the county consisted of 237,321 acres of farmland, but by 1969 that number had decreased to 215,055 due to changes to urban boundaries.[3]

In response to concerns that urban growth was getting out of hand, the Oregon Legislature passed Senate Bill 10 in 1969. The bill required local county governments to develop land use zoning initiatives and implement them by the deadline of December 21, 1971. It also gave the governor the authority to take control over the planning of any county that had not been completely zoned or had not made significant progress by 1971. SB 10 was vague in its wording, and did not give specific directions for the county to follow. Additionally, the state did not provide funding for counties and lacked effective enforcement to ensure that regulations were being met. By 1971 very few counties had actually met the deadline and the governor's office lacked the administrative ability or the political will to take over the zoning of non-compliant counties. Despite the flaws in implementation of Senate Bill 10, with 55 percent of state voter support as seen in a referendum, it succeeded in paving the way for public support of Senate Bill 100.[4]

Senate Bill 100

Senate Bill 100 was signed into law during the 1973 legislative session. SB 100 improved on SB 10 by allocating a more significant role to the state government in local land use planning and regulation. SB 100 mandated that all cities, regional agencies, and all other local agencies that had planning authority prepare comprehensive land use plans consistent with state goals. Comprehensive plans had to be accepted by an elected governing body, such as city councils or county commission boards, in order to ensure accountability by the public. To oversee and enforce this process, SB 100 created the Land Conservation and Development Committee (LCDC; hereafter, the Committee) as well as the Department of Land Conservation and Development (DLCD). The governor was mandated to appoint seven members to the Committee, whose task it was to review and recommend areas in acute need of conservation and land-use planning. During the process of implementing SB 100, the Committee appointed a Citizen Involvement Advisory Committee to provide for public input and keep projects on track and working in the appropriate direction.[5]

SB 100 succeeded where SB 10 did not due to its strict enforcement of objectives through the Committee. The Committee was required to determine if local plans were in compliance with statewide goals. This process began when the local comprehensive plan was sent to the DLCD, where it was reviewed before being sent to interested parties for comments, and then finally submitted to the Committee for a public hearing. At the public hearing the Committee would determine whether or not the plan complied with the established guidelines laid out by SB 100. Until a plan was accepted, the state retained all authority over local land use decisions. In the event that a county or city was noncompliant or did not receive acknowledgement by January 1, 1976, the Committee had the authority to write its own land use comprehensive plan for the locality, and charge the local government with the cost of its creation. The Committee also had the authority in these scenarios to suspend or permit development, strong-arming the local government into compliance.[6]

Land Use Planning as Applied at Zena

The land use plan created for Polk County in 1974 states that its main goals are to create policies for the conservation and development of community resources, with an emphasis on mitigating future growth. The growing population during the early 1960s presented new concerns for urban expansion, and prompted decreases in available agricultural land. The number of farms in Polk County decreased from over 1300 to less than 1100 between 1959 and 1969. The land use plan outlines four major issues: (1) The current land use pattern is disorganized and unsightly, which is inefficient and should be improved for better systematic planning. (2) It is economically beneficial to have a streamlined plan. (3) A disorganized plan utilizes more resources than necessary. (4) There are less urban-rural conflicts when a more effective and efficient land use plan is in place.

The Polk County Comprehensive Plan (PCCP) attempted to address these issues by dictating land use regulations for urban industrial and commercial areas, as well as rural agriculture and forest properties. The Zena property lies within the agricultural

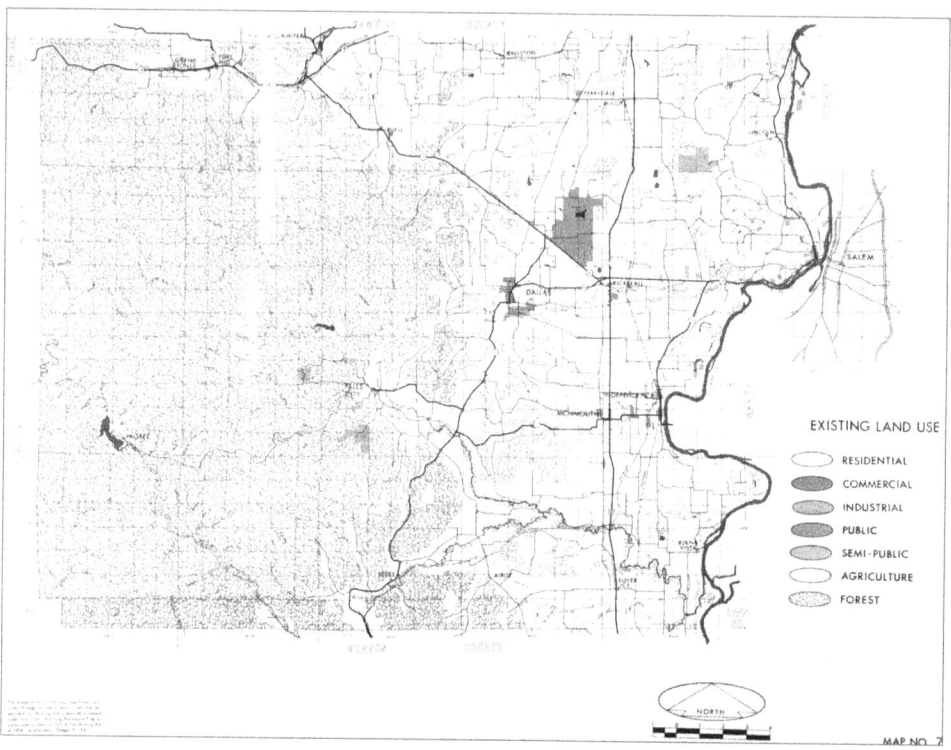

11. Polk county land use, past and planned. *Polk County Comprehensive Plan (June, 1974)*

area designation, an area which includes 179,500 acres and is distinguished by fertile soil and larger property sizes. As stated in the PCCP, the designation of these lands was to maintain "the agricultural economy for the county by strictly limiting nonagricultural development in the area." The only other development allowed is that which is essential to the farming community, including churches, schools, and parks. Through such regulations, the PCCP sought to maintain agricultural areas like Zena both for direct benefits, such as food production and employment, and indirect benefits, including scenic appeal and lush natural space.

Reactions and Responses

Some Zena area residents, while supporting land use planning, argue that SB 100 has been poorly designed and negatively impacts their livelihood. Anne Walton, the property owner adjacent to Willamette University's farm and forest, notes that SB 100 placed

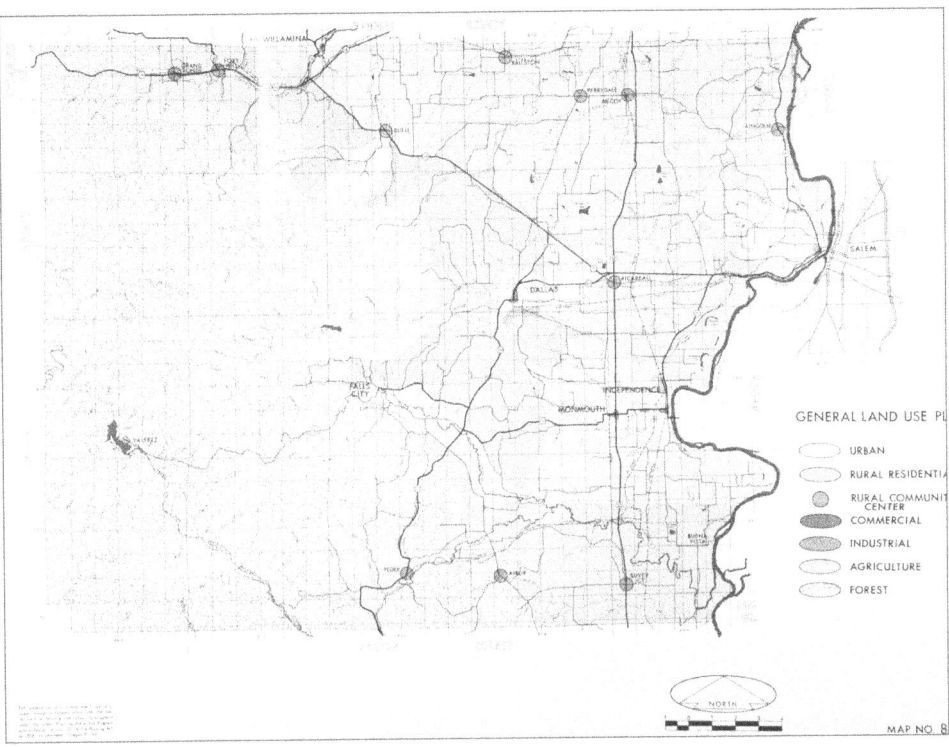

extreme limitations on how larger rural properties may be used. The implementation of SB 100 has essentially prohibited further subdivision of her property; also, additional buildings cannot be erected without going through an intensive permitting process. The idea of preventing subdivision is generally accepted, as the water supply would likely not support such growth. However, from Walton's perspective, the details of the bill seem to expose the uninformed nature of the lawmakers. While their aim to keep families on the farm is beneficial to the county, Walton notes the building restrictions have both limited farm productivity and also the choices she can make about her land. Having family ties to the land grants a certain perspective, and a feeling of a sense of place with the property. Walton noted that, ideally, she would build the farm house on a different ridgeline, which would allow a more inclusive view of the valley; however, that would conflict with zoning regulations. She comments that after SB 100, property ownership feels more like managing an estate than running a farm.[7]

Bob Feldman, who lives near Zena, outlines a different perspective on SB 100 and land use planning. Feldman purchased his property in 1966 with the intent of starting a sheep ranch, which he did for a while until he grew tired of dealing with the potent combination of thieves, dogs, and coyotes. In 1976, Feldman switched to growing Christmas trees, which he has been doing ever since. When asked how he felt about Senate Bill 100, Mr. Feldman responded "If society wants to preserve my land, they should buy it from me." On February 15, 1989, Bob and Kathleen Feldman wrote a letter that they sent to thirty state senators and sixty state representatives, expressing their distaste for Senate Bill 100 and land use in its current state. In the letter, the Feldmans recounted their experience of a public meeting in Salem they attended in 1989, which was held by the DLCD as a way for citizens to have a final voice on the matter of land use planning. According to the letter, approximately 500 people were in attendance, out of whom 110 testified with only two actually in favor. They continued by describing the attitudes of the board members, describing them as uninterested and apathetic to their plights: "We felt personally insulted and ignored by the committee's performance. They were eating and wandering around while people were attempting to defend their very basic right of property ownership." What the Committee and the urban supporters of SB 100 did not understand was that a farmer's livelihood was his property. According to Feldman, if a farmer misused a piece of his land, he put his own welfare in jeopardy and therefore did not need a "faceless bureaucrat" to tell him how to care for his land.[8]

Perspectives on Land Use Planning and Sense of Place

Land use planning and its implementation in Oregon was a pioneer process, and it was paramount in preventing the rural agricultural land from being developed and taken over by urban sprawl. But ultimately one's perspective on the land and how it should be used and divided depends on one's sense of place and connection to the land. This differs for each person, as farmers, bureaucrats, and urban residents each have different needs and appreciations of the land. The process and results of land use planning impact people's sense of place and connectivity with the land, as some want to maintain their rural history with the land while others want to gain a rural connection. Motivation by urban dwellers to develop individual senses of place in the rural landscapes of the Willamette Valley ultimately created the very development they wished to escape. The implementation of land use planning has mitigated many of the issues regarding urban sprawl, yet has also created some disenchantment with the state legislature and land use agencies.

At Zena specifically and the area around it, individual sense of place is deeply rooted in the utilization of the land, which has sometimes come into conflict with the regulations set down by SB 100. As suggested by Anne Walton's comments, the historical aspect creates strong ties and places incredible value on the land, but zoning has severely impacted her use of the land and its value. For Bob Feldman, the entire process of land use planning has been alienating and insulting. For Willamette University, the sense of place attached to the land at Zena is different for each person who spends time there, and is in a way influenced by the land use regulations. The Zena property abides by the regulations set in place by SB 100 handily, as its uses are primarily for educational agriculture and forestry, as stipulated by the more rigorous guidelines set by the conservation easement coupled with the University's purchase of the land. And so while land use planning has in some ways frustrated how other area residents connect to their land, for Willamette University, land use planning has fit nicely with the school's use, vision, and sense of Zena as an aesthetic and educational place.

Notes

1. Adler, *Oregon Plans*, 24; Medler and Mushkatel, "Urban-Rural Class Conflict in Oregon Land Use Planning," 349.

2. Abbott et al., *Planning the Oregon Way*, xi, 6; Walker and Hurley, *Planning Paradise*, 6; Leonard, *Managing Oregon's Growth*, 4.

3. Polk County, *Comprehensive Plan, 1974*, 29, 36.

4. Abbott et al., *Planning the Oregon Way*, xiii; Walker and Hurley, *Planning Paradise*, 44.

5. MacPherson and Paulus, "Senate Bill 100," 414–21.

6. Knaap and Nelson, *The Regulated Landscape*.

7. Anne Walton, in discussion with Kyle Carboni, 28 February 2012, and Morgan Gratz-Weiser, 27 October 2012.

8. Bob Feldman, in discussion with Kyle Carboni, 27 March 2012, and Morgan Gratz-Weiser, 31 October 2012.

10

Sarah Deumling at Zena Forest

Lauren Henken

Encompassed by magnificent Douglas firs, Oregon white oaks and broadleaf maples, it is easy to replace the commotion of twenty-first century life with the tranquility and empowerment of the forests and meadows, and to get lost in the timeless woods known as Zena Forest. The soft spring air, warmed by the timid Oregon sun, rustles the oaks that surround Sarah Deumling's rust-red house with pine-green trim. She sits on the porch, overlooking her grassy yard to the forest beyond, lost in memories of what has taken place here over the years. She is dressed in jeans, a gray fleece, and hiking boots, all traced with mud and well worn from her days of labor. A native Oregonian and the former owner of Willamette's Zena Farm, Sarah Deumling is humble and lighthearted and has an immense love for her children, her gardens, and her forests.

Having grown up on a farm in the Willamette Valley, Sarah says she has always felt a connection to the land here and a love for working with it. That connection and sense of place, she explains, has only grown stronger as she has experienced more of Zena Forest and gained

12. Sarah Deumling and her dog, Henry. *Courtesy of Lauren Henken*

a deeper understanding of the land and the flora and fauna that flourish here. A sense of place is an individual's unique relationship to the land and the space that is created there. Such a relationship is established through sensory interactions with the land, connection to the land built on experiences that occur there, and an understanding of the land, which is developed through the value judgments of that place. An individual's sense of place is also influenced by the history of a place, the possibilities for the future, and the witness of the passing of time. Consequently, an individual's sense of place is as dynamic and fluid as the seasons and the place itself. The narrative of Sarah Deumling's acquisition, stewardship, and love for Zena Forest exemplifies a sense of place that has continually changed as her interactions and relationship with, and knowledge, understanding, and valuing of Zena Forest have grown and transformed.

"I had no idea if I could do it or would want to do it," Sarah reflects on her first thoughts of managing Zena Forest when her husband, Dieter Deumling, passed away from lymphoma in the fall of 1996. At that moment Sarah, who had previously been a grade school teacher, was presented with two options: leave Zena Forest, take her children with her, and try to make a living somewhere else, or attempt to learn how to manage a forest. When Sarah reflected on the discussion with her children about their options, she recalled that their response of, "We do not have to move, do we?" as the writing on the wall, and she told herself, "I think you are going to try forestry."[1]

Although Sarah now lives in the same valley in which she grew up, her path to management and ownership of Zena Forest took her to Germany and back several times. During her junior year at Molalla High School, Sarah's family hosted a German exchange student, Apollonia Heisenberg, and the two became close friends and still speak regularly. "We are like sisters," explains Sarah. After Sarah finished high school, Apollonia's family invited her to visit them in Germany. It was there that Sarah met her future husband, Dieter, when the two girls were headed to an end of the year party. Sarah recalls the scene: "My friend invited two boys, who both thought they were going with her. But one of them was suppose to take me, but she didn't say that! And the one who was supposed to take me was really miffed . . . but he's the one I ended up marrying eight years later."

Dieter had always been fascinated with the United States and especially the West. He had listened to Voice of America growing up, through which he learned English, and watched "American television because he said it was the best way to get to know American culture." The two moved back to the United States and got married in 1970. Sarah had a job teaching English on a Navajo reservation and Dieter went to Northern Arizona University for his master's degree in American Studies and History of the American West. Dieter never actually studied forestry, but one day in 1980 he got a call from Count Hermann Hatzfeldt, the owner of a large forestry business in Germany and a friend Dieter's previous employer, who offered Dieter a job.

Count Hermann Hatzfeldt was a foreword-thinking businessman and motorcycle-loving environmentalist. The Hatzfeldt family has owned forestry land in Germany for hundreds of years, yet when Hermann inherited the land and family forestry business, it came as a surprise. Hermann had an older brother, Christian, who would have ordinarily taken on the responsibility of ownership; however, the family felt he was too much of "a playboy," and considered Hermann to be more responsible and better suited for the job. Although unplanned, the Count accepted the career "with grace and commitment," described Sarah. Hermann took the family business in a new direction, away from single species plantations to more diversified crop rotations, with a greater respect for nature and consideration for ecological concerns. This change in values was instigated by an increased awareness of environmental degradation as well as the quick succession of two severe windstorms, which ripped across the Hatzfeldt plantations and destroyed the single specie rows while the mixed specie stands survived. As Sarah remembered, those disasters made the Count think, "There is something wrong with the way we are doing forestry." Accordingly, Count Hatzfeldt redirected the family forestry practices and diversified his assets by purchasing woodlands in the United States. This, Sarah explained, required "someone who was fairly familiar with both countries, and languages, and mindsets, and economies," and Dieter had such qualifications. The young couple thought that it might be a valuable opportunity, so they packed up their family of two children, Reuben and Katherine, and moved to Germany for eight years. During this time, Dieter's job evolved into the self-titled position, Environmental Minister of

Hatzfeldt. The job involved a lot of public relation work, including tours of the forests, lobbying for fewer emissions, and developing library archives on the subject of environmental threats to forests.

Things were great over there in Germany, reflects Sarah. However, she remembers sensing the importance of being grounded in a physical place, and began to have concerns that her children—the older two born in the United States, and the younger two born in Germany—might lack connection to a place and a place to call home. As Sarah recalled, "I got the feeling that these children need to know where home is. So either we need to decide that we are going to stay in Germany or we have to decide that we are going to go back to Oregon." This was during the 1980s, when acid rain was becoming a more pronounced issue for European foresters, and there were "a lot of questions about the long-term health of the German forests and the European forests." Subsequently Dieter, who loved Oregon and wanted to return there, and knowing that the Count was keen on diversifying his assets, "convinced the Germans that the healthiest forests in the Northern hemisphere were in Oregon and Washington," that those forests would be most productive, and that it would be wise to purchase woodlands there.

For several subsequent summers, the Deumling family ventured to Oregon to visit and, as Sarah described, "snoop around for what was for sale in the Willamette Valley." Dieter found the first 400 acres of this Zena property in the summer of 1983 and purchased the land the following summer. After the German delegates traveled to Oregon and approved the initial site, each summer thereafter Dieter found more tracts of land until Count Hatzfeldt owned close to 2,000 acres. As his share of Oregon forests grew, the Count hired an Oregon-based forestry company to manage it for him, but their practices were traditional to forestry management across the United States, which emphasized large profits through clear-cutting, chemical use, and the use of heavy equipment. The Count did not practice such methods in Germany, and he and his advisors were unsatisfied with the US forestry company's lack of ecological consideration. Subsequently, the Deumlings proposed that they return to Oregon and manage the Count's forests—which, as Sarah explained, was their hope and intention all along, "so it worked out."

In the heart of Zena Forest, the Deumlings were allotted 40 acres as personal property for their home and, as the years progressed, a couple of gardens, a barn, an outdoor stage, and a cemetery. When they returned, Dieter had sole responsibility for the care and management of the forests. "I just did the children and the household and the garden and whatnot. I was not involved in the forestry part," recalls Sarah. Nonetheless, she grew greatly attached to the land of Zena Forest, appreciating it for its intrinsic beauty and its ability to sustain her family through her garden and the sheep she raised. At this time, her sense of place at Zena Forest was founded on her understanding of the land through her garden and through the interactions she and her children experienced there; the forest and its value constituted only a vague imprint on her sense of place. "I did not know the forest at all," Sarah reminisces. "Once in a while Dieter would take me on a walk and say, 'Come look at something,' but that was about it." In truth, she remembers at times being quite critical of her husband and his management decisions, for she did not fully understand the challenge of balancing timber profits with forest restoration. "I was pretty clueless. We had a better marriage when we each did our own thing, than if I messed with what he was doing. I tended to be critical. . . . I said, 'You are not going to build another road are you? There are beautiful things where you want to put that road.' Now I realize if you are going to practice forestry, you need to have roads. You have to make some compromises. Or else it becomes a park, which is great, but you cannot live off of a park."

This shift from a more limited sense of place to a more profound and complex connection to a place and its multitude of uses and values can only come about from experiencing that place and fostering a personal relationship with the place. As Sarah describes, a sense of place comes from "just being there; there isn't any substitute." Sarah's relationship and sense of Zena Forest was altered in several ways when Dieter passed away in 1996. Daily occurrences with her husband became memories connected to a certain physical localities, and she became more aware of the changes and interactions that took place throughout Zena Forest as they reminded her of him. Yet, his passing left her with an amazing opportunity to manage Zena Forest, a chance that has created for her a dynamic and growing sense of place through her increased understanding, appreciation, and love for Zena Forest.

When Dieter passed away, Sarah's two youngest boys were 12 and 14, and having grown up among the forests and meadows, they loved Zena Forest and did not want to move. For that reason, Sarah decided that she would try forestry. "I said I would try for three years. I would give myself three years." Supporting her decision, the Count sent Dr. Franz Straubinger, the head forester for the Hatzfeldt forests, to the Willamette Valley once a year for a decade to teach Sarah about forestry and to assist her in managing Zena Forest. Sarah recalled that Dieter had not gotten along well with Dr. Franz Straubinger because he preferred to run his own show. "But I could not run my own show because I did not know enough . . . so the head forester came and spent two months teaching me. We went out every day to the woods. It was very intensive, hands on. 'This is what you need to know.' We practiced planting trees, laying out roads, and all sorts of stuff. He was good." Some summers and falls, Count Hatzfeldt, Dr. Straubinger's wife, and other German loggers and managers would accompany Dr. Straubinger to Oregon. "So there was a German contingent here every October," Sarah reminisces. During those months, her children referred to their home as "The Hunting Lodge" because of the influx of Germans who came to live there as well as hunt, check up on their forests, give Sarah tutorials, and plan the budget for the next year. Sarah recalls that those times were rewarding and fun, but also a lot of work, because it was just her.

Dr. Franz Straubinger taught Sarah about *naturgemässe waldwirtschaft*, a German forestry term which translates to "close-to-nature forestry"—a respect for the interplay of species and an understanding that the land knows what will prosper there. This sustainable model of forestry has been at times misunderstood as synonymous with German forestry in general, but there are other models of German forestry that do not place such great value on the resilience and overall health of the forest. However, during the eighteenth century it was the Germans who first began to develop a systematic approach to forestry and the cultivation of trees—the discipline now known as silviculture. Silviculture is both an art and a science that aims to understand and control the growth, composition, health, and overall quality of the forests. Silviculture first appeared in the United States in the late nineteenth century, and in the early twentieth century began to resonate with foresters as concern over resource conservation grew. Silviculture considers site-specific biological and social science concerns when developing forest management strategies in order to improve forest stands while earning a profit; what the Germans referred to as *dauerwald*, meaning "permanent forest." It was this dual management strategy of stand restoration through reduced compaction, limited chemical use, and mixed stand maintenance, while simultaneously earning a profit, that Count Hatzfeldt chose to utilize throughout his German woodlands—and the strategy which Sarah Deumling also chose to adopt and implement at Zena Forest.

During the mid-1990s, Count Hatzfeldt and the Germans continued to diversify their assets and began investing "in large tracks of forest land in what used to be East Germany before the Wall went down." Sarah explains that since this land was relatively inexpensive and closer to home, their focus turned to it, and Zena Forest "became like a fun little toy, but it was not very sensible." As a result, they decided to sell Zena. Sarah remembers the initial shock she felt at hearing this news: "My family felt like it was theirs, even though it was not. Because we lived at Zena Forest, and took care of it, and knew it better than anyone else." The thought of the possible loss of what she, Dieter, and their family had worked so hard to create, and the loss of a place to which they had grown so deeply attached, was terrible. However, she had faith in the Count, who Sarah describes as an extraordinary man and "a sort of visionary" concerned with "the long-term health of the forests." So she thought, "We will see what happens."

At this time, Zena Forest was over 2,100 acres, which the Germans initially tried to sell outright for eight million dollars. They were looking for a conservation buyer who would manage the forest in a similar fashion to themselves, without the use of clear-cuts and chemicals, and limited compaction of the soil. However, they were unsuccessful in finding a client who was interested in land with such astringent limitations. Apprehensive about the fate of Zena Forest, Sarah contacted an acquaintance at the Trust for Public Land (TPL). As Sarah explains, the TPL "does not own land, but brokers deals to save land of high conservation value." The TPL turned out to be a "terrific partner" because they have strong connections to the Bonneville Power Administration (BPA). As a result of environmental damage caused by hydroelectric dams in the Pacific Northwest, BPA has federal mitigation requirements to save high conservation resource land. Zena Forest qualified as high conservation land because it included large areas of endangered oak savanna and oak woodland. About 40 percent of the trees in Zena

Forest are Oregon white oaks. This relative abundance of Oregon white oaks stands in stark contrast to the Willamette Valley's 3 percent of its original white oak population. Historically, Oregon white oak was not considered a merchantable tree. As Sarah elaborated, "In Oregon we have only sold conifer and primarily Douglas fir, which is a fast growing, magnificent tree that makes the best building material in the world . . . and most of the oak has gone to farming, development, and now vineyards. So oak habitat is very important and a much diminished ecosystem."

Consequently, the BPA was interested in purchasing a conservation easement, and for the first time in the Willamette Valley, the BPA was interested in a working forest easement. This meant that Sarah could still practice commercial forestry, provided that she also work towards maintaining and enhancing the conservation value. This, as Sarah expresses, "was so wonderful for us because it is exactly what we have always believed in. That the two really are the same"—referring to the relationship between forest restoration and timber profits. Sarah elaborates further, "If your ecology is in order, you are going to have a healthier, more vital and more valuable forest in the long term." Accordingly, the conservation easement suited both the Germans' and the Deumlings' philosophy.

Within the 2,100 acres of Zena Forest, there were three noncontiguous parcels, between 100 and 200 acres each. Before TPL took over the negotiations, the Germans decided to sell those three parcels without restrictions because they needed the money to move forward with their plans to purchase German woodlands. They received high offers from vineyard owners, but they were reluctant to sell the land to them, still wanting the parcels to remain forested. For that reason, the Germans sold it to another timber company, which was headquartered in Arkansas but had mills in Oregon. As Sarah comments on the fate of the parcels with disappointment, "All three were clear-cut the next summer. Everything's gone. Two of them are now vineyards." As this series of events unfolded, Sarah grew more concerned about the possible fate of Zena Forest depending on who purchased the land. As Sarah describes her thoughts at the time, "I have forty acres, and this whole thing is for sale, and I do not want to be stuck with forty acres in the middle of a clear-cut. That would just be too awful." After considering her options, Sarah decided to borrow enough inheritance from her mother—who at the time was 94—to buy 130 acres of land to the north and south of her house, which increased her private land to 170 acres. She hoped that regardless of what the future might hold for the rest of Zena Forest, her home and her family's lifestyle would be protected by the Douglas firs, Oregon oaks, and broadleaf maples that she and her family had grown so accustomed to.

About this time in 2007, the conservation easement was confirmed, and the Germans were paid five million dollars in development rights. The price of the conservation easement was determined in part because of the quality of the land, in part because the Germans gave up the construction rights of the property, and in part because the easement restricted how the forest could be treated and timber harvested. Accordingly, the value of Zena Forest fell significantly, thus TPL agreed to pay an additional three million to the Germans. "So the Trust for Public Land owns this place, even though they do not own land generally," explained Sarah. However, they took on the risk with the hope that at such a low price, they would be able to find a conservation buyer without much trouble.

Unfortunately for TPL, with the conservation easement Zena Forest proved difficult to sell. While TPL struggled with the sale, the BPA asked Sarah if she would like an easement on her land as well. Which, as Sarah describes, "I hadn't even thought of. For some reason, I just spaced it. . . . Well of course we would!" So she replied to the BPA with affirmation, and in turn received over a million dollars, which she recalled was an amount unheard of to her. Owing to the combination of the decreased price tag and the large sum of money she just received, Sarah began to wonder, "how far are we now from being able to buy this?" As she explains, "well we were still very far from a normal human being's perspective. So we thought about a community forest. We thought about making it a nonprofit and we thought about all our rich friends, none of whom were interested." That being the case, Sarah called up Willamette University because she knew the Dempsey Professor of Environmental Policy Joe Bowersox quite well, for he and fellow Willamette professors Karen Arabas and Susan Kephart had brought classes out to Zena Forest for years.

Sarah described to Joe—whom she knew best—the situation and the lowered price, and asked him if Willamette University would be interested in purchasing the land fifty-fifty and managing it together. "He just went for it hook, line, and sinker," as Sarah describes his response. However, the Willamette trustees were still not interested. While searching for a way to purchase Zena Forest, the Deumlings realized that they had one disposable asset: a beach house that had been in the family for a long time. Sarah explained that, on a whim, after the family all agreed, they "put it out to friends and relations that the beach house was for sale . . . and in two weeks I had a buyer for the price I asked. So here we were; we were 800,000 dollars short." Simultaneously, the TPL—recognizing the difficulty of finding an appropriate buyer—lowered their asking price and invited Sarah to sign a purchase agreement. However, not wanting to go into debt, Sarah stated that she would only sign the purchase agreement if the TPL inserted a clause allowing her to sell two tax lots equaling 305 acres. The TPL approved the clause, recollected Sarah, "So I signed this purchase agreement, and then I called up Joe Bowersox and said, 'I just signed a purchase agreement,' and I could hear that he nearly dropped the phone. I said, 'but you can have 305 acres and the house for 800,000 dollars.'" A couple days later, in the fall of 2008, Joe called back and responded that they had a deal, and those two tax lots became Willamette University's and are now known as Zena Forest and Farm. So it all worked out and "everybody was happy," Sarah reflects. "It is a magical story."

This final agreement again altered Sarah's relationship to Zena Forest: from a connection founded on care and stewardship of the land to a connection through her ownership, complete responsibility, and freedom to manage Zena Forest as she saw fit. Consequently, a new sense of place began to grow as Sarah stepped into ownership, took what she had learned from the Germans about *naturgemässe waldwirtschaft*, which seemed "infinitely sensible to me," and made their dual-management system more uniquely her own. She has since redefined her own vision for the future of Zena Forest: "That I will leave the forest better than I found it. More resilient than I found it and a healthier ecosystem." Sarah Deumling's story of her acquisition, stewardship, and immense love for Zena Forest illustrates that a sense of place is as fluid and dynamic as the seasons experienced there. Sarah's sense of place at Zena Forest has continually developed as her interaction and relationship to the land has been redefined, as her knowledge and understanding of the forests has become more profound, and as her value and love of Zena Forest has grown. As Sarah reminisced about her first thoughts after Zena Forest became hers, "I can do anything I want with it. It felt very much our own now. I mean I wake up every morning and have to pinch myself. It is all yours! You get to take care of it. Protect it." Although everyone can experience a sense of place, one as complex, profound and fluid as Sarah Deumling's sense of place at Zena Forest is rare and, as she concludes, comes from "just being there . . . opening up your eyes and paying attention."

Notes

1. Sarah Deumling, in discussion with Lauren Henken, 9 March and 2 April 2012.

11

A Bureaucratic Sense of Place

Philip Colburn with Kevin Bernstein

How does a university come to own a forest? Simple: someone else pays for it. In the case of Willamette's purchase of Zena, the process was facilitated by what is known as a "conservation easement." Entities needing to make reparations (typically because of environmental destruction) will earn market-based credits for habitat preservation by purchasing a piece of property and placing the property under a conservation easement, having another party manage the land under requirements dictated by law. The circumstances that led up to the purchase tell a unique story that could have only happened due to the cooperation of many different agencies with individual goals that simultaneously converged on Zena Forest.

Environmental historians like Samuel Hays and Hal Rothman speak of environmental dialogue changing from the early 1900s from a view about aesthetics and health to a scientifically dominated narrative. That narrative focuses on the peak of environmentalism in the 1960s and 1970s, when federal laws were enacted at an incredible rate, furthering conservation in society. Environmentalism encountered strong opposition with the election of Ronald Reagan in 1980, but laws that had already been enacted by earlier presidents like Richard Nixon and Jimmy Carter still have had measureable effects, which furthered the cause of environmentalism.1

This chapter tells the story of the conservation easement at Zena as a way of exploring the diffusion of preservationist values in American government and culture since the 1960s. In addition to institutional histories and studies produced by both government and private agencies, this chapter draws on interviews with a variety of important actors: Karl Weist from the Oregon office of the Northwest Power Conservation Council, Michael Pope (formerly) and Laura Tesler (currently) of the Oregon Department of Fish and Wildlife, and professor Joe Bowersox, chair of the Environmental and Earth Sciences department at Willamette University. This narrative of the creation of the conservation easement and the forestry practices guided under the law suggests that environmental stewardship does not always require an intimate sense of place. In the case of Zena, two agencies (the Bonneville Power Administration and the Trust for Public Land) with little close contact to Zena, sought to protect that specific place. These agencies' "bureaucratic sense of place" created the framework within which Zena became both a laboratory for sustainability and ecological learning, and an example of the spread of environmentalism in the late twentieth and early twenty-first centuries.

A Brief History of Bureaucratic Environmentalism

Efforts to protect the nonhuman natural world stretch far back into American history, predating even 1872, when Yellowstone became the first national park. But the environmental movement as we know it came into being in the 1960s, when environmental damages became a national issue. Rachel Carson's 1962 seminal exposé, *Silent Spring*, revealed the ecological harm of dichloro-diphenyl-trichloroethane (DDT), and the 1969 Santa Barbara oil spill brought the horror of immense ecological damage into the public's knowledge through media attention. Public outcry led to wide-ranging environmental legislation, often with bipartisan support. The National Environmental Policy Act (NEPA) of 1969 transformed the federal government's job from conservation of natural resources to protection of the wilderness. The bill creating the Environmental Protection Agency (EPA) was signed into law in December of 1970 by Richard Nixon, adding the agency to the executive branch for the effective enforcement of environmental policy. This agency was in charge of developing the Endangered Species Act (ESA), first created in 1966 under the name of the Endangered Species Preservation Act. The Act evolved to become the Endangered Species Act in 1973, and gave the EPA jurisdiction over protecting the habitat of endangered and threatened species.[2]

The story of modern environmentalism merged with the history of Zena as a result of public uproar over the disappearance of salmon due to damming of the Columbia River and its tributaries. The Bonneville Power Administration (BPA) is a governmental organization that provides power to the Pacific Northwest primarily from hydroelectric dams. The initial investment in dams in the 1930s on the Columbia River and its tributaries supported two initial goals: providing cheap clean power and preventing flooding through the Willamette Valley. The construction of the dams caused a decline in salmon runs in two ways: by creating physical barriers to the mature salmon's journey to spawning locations and by killing the salmon smolt on their maiden voyage to the ocean. While the Pacific salmon were not a listed species under the ESA until the late 1990s, regulations to protect salmon began a decade before that. In 1980 Jimmy Carter signed into law the Northwest Power Act (NWPA), empowering a new council of two delegates each from Oregon, Washington, Idaho and Montana to create programs to be implemented by the BPA. The measures in the bill were specifically designed with three goals in mind: first, produce a plan for the Northwest for an adequate, efficient, economical, and reliable power supply; second, create a plan to protect, mitigate, and enhance fish and wildlife that have been affected by hydropower dams in the Columbia River Basin; third, inform and involve the public about regional energy issues. To enforce these measures, the NWPA also created the Northwest Power and Conservation Council (NPCC).[3]

Among the more significant environmental provisions of the NWPA is the law's requirement for an assessment of the habitat lost by the creation of dams. The BPA is obligated to preserve the same amount of habitat as was destroyed by the construction of the dams. To do so, the BPA usually purchases land outright for preservation, called fee-title purchases, but in the case of Zena (and elsewhere in the Willamette Valley), they arranged for a conservation easement that also serves the purpose of habitat preservation. Conservation easements are funded and donated by the BPA and remain in private ownership at a greatly reduced price. The BPA is legally responsible for enforcement of the conservation easement regulations and grants guardianship duties, with development restrictions, to the landowner.[4]

Through Zena, the BPA earned habitat conservation credits compensating for the habitat losses created by the dams at Detroit, Oregon. Detroit Dam and Big Cliff Dam, completed in 1953, are located on the North Santiam River, which feeds into the Willamette

Valley Basin. The Zena purchase in 2008, totaling 1,797 acres, was the largest conservation easement the BPA had purchased. Zena is home to a diverse set of biomes—from oak savanna to old growth forest—and this diversity was a key criterion in selecting the location as a candidate for conservation. Of the nearly 1,800 acres in the easement, 305 are managed by Willamette University. The remaining land is in the ownership of Sarah Deumling, who practices Forest Stewardship Council–certified sustainable forestry.[5]

The story of Zena involves many agencies that never really had any particular connection to Zena and yet worked together to protect Zena into perpetuity. This presents a new perspective on sense of place, which will be explored through the analysis of the two agencies most important to the conservation easement at Zena: the Bonneville Power Administration (BPA) and the Trust for Public Land (TPL).

Bonneville Power Administration

Congress established the BPA in 1937 to make available and distribute the power generated from the Bonneville dam to the people of the Northwest. The BPA was actively engaged in an idea of conservation that stemmed from conserving natural resources for "wise" use. The BPA accomplished this goal through a system of transmission lines for public and private power that connect 31 dams and one nuclear power plant.[6]

In 1980 the BPA was legally obligated to add to its mission statement the goal of "Mitigation of the Federal Columbia River Power System's impacts on fish and wildlife." In the Willamette Valley alone, 26,537 acres have been impacted by the hydroelectric systems managed by the BPA. In 2010, in conjunction with the Oregon Department of Fish and Wildlife (ODFW) and the Northwest Power and Conservation Council (NPCC), the BPA created a new policy that conserves habitat through conservation easements based on overall acreage. As of October 2010, 9,657 acres were protected by conservation easements under the NWPA. For the remaining 16,880 acres to be mitigated, the NPCC has received 117 million dollars from the BPA for completion by 2025. Conservation locations are selected by the ODFW, which then request funds from the NPCC to set up conservation easements. The BPA maintains a role in the purchase by holding public review on lands before the purchase is finalized and contracting yearly Forest Stewardship Council (FSC) inspections to the Scientific Certification Service (SCS). It has also started a pilot program of a five-year review process on existing conservation easements. This allows for the BPA to have a relatively hands-off approach and maintain focus on electricity generation.[7]

This particular approach sometimes appears as a lack of genuine attention to the details of environmental preservation at Zena. During the first five-year review of the conservation easement, environmental science professors Joe Bowersox and Karen Arabas conferred with the BPA about the progress of conservation activity at Zena. Bowersox noted that in their review, the BPA seemed disconnected with the Zena property:

> [The BPA explained] "We are doing this checklist because we want to see if you are fulfilling the requirements of the conservation easement. . . ." They started asking these sort of questions, [and Bowersox and Arabas responded] "Well, you know we have been answering these questions for the auditors that have come from the Scientific Certification Service every year and I assume that you have been getting these reports." There were these sort of blank faces, and they look at each other and go, "Do you know where that report is?"

The BPA, it seems, had never seen the SCS reports, even though the BPA had commissioned those reports every year, as the inspections are required for the conservation easement to maintain tax-exempt status. This reveals some of the limits to the BPA's particular connection to Zena.[8]

While the BPA may not have a strong sense of place at Zena property, the agency nevertheless plays an important preservationist role. The BPA works with many different organizations—the NPCC, the SCS, ODFW and TPL—to produce effective conservation easements like those we see at Zena forest. While the narrative of the rest of this book often focuses on close relationships between people and the land, the story of the BPA reveals that a personal, intimate sense of place is not a prerequisite for participating in preservation.

The Trust for Public Land

At the Trust for Public Land, in contrast to the above, sense of place goes to the roots of their organization. The TPL was created in 1972. Inspired by the first Earth Day, groups of lawyers, real-estate professionals, and finance experts joined together to create an organization the express goal of which was conservation. The groups formed an organization as a means to preserve habitats that members had enjoyed throughout their lives, using their sense of place as an inspiration to protect habitats so that others would be able to develop their own connections to the nonhuman natural world. The TPL is a nonprofit, nongovernmental organization (NGO) that gets funding from donations, foundations and government initiatives. Currently, the TPL has three working initiatives: (1) conserving wilderness to address the effects of climate change and to preserve clean drinking water, (2) ensuring access to public parks, and (3) protecting farms, ranches, forests and other working lands to support land-based livelihoods while preserving environmental benefits. To protect identified locations, the TPL works as a middleman, purchasing lands in need of protection and setting up conservation easements with two parties, one who earns the conservation easement credit through funding most of the purchase and the other who manages the property. The TPL is thus financed on a revolving door type policy, in which sale of old easements helps to pay for new purchases and development of easements.[9]

The TPL purchased the Zena Forest property in 2006. The land had already held an FSC certification since 1998, which was put on the land by its owner, Count Hatzfeldt and practiced by Sarah Deumling. In the sale to the TPL, Count Hatzfeldt stipulated that the conservation easement must remain FSC certified, wishing to avoid the clear-cut fate that had befallen some of his other former properties. TPL then began looking for a buyer, a process that took longer than expected—and the longer it took, the lower the price fell. This reduction in price facilitated the purchase by the BPA and Sarah Deumling. The BPA paid 6,485,000 dollars for the conservation easement, giving the BPA the environmental credits, and the property rights were sold to Sarah Deumling at a price of 3,000,000 dollars. The BPA remained in control of the environmental credits, while Sarah Deumling became the property rights holder. Willamette University wanted to use the forest as carbon credits, but the BPA would not allow this credit, as environmental credits cannot be "double dipped." The BPA remains the sole environmental credit holder of the Zena property.[10]

Conservation Easement and Habitat Protection at Zena

When Willamette University purchased 305 acres of Zena Forest from Sarah Deumling in 2008, it became the first educational institution under the BPA's program to own a conservation easement. Under the conservation easement, Willamette is allowed to use the land for educational purposes as long as it meets the goals of the easement. Should a person wonder how to achieve these objectives, the Environmental and Earth Science department at Willamette can supply the FSC guide to Sustainable Forestry at Zena as drafted by Trout Mountain Forestry (TMF), a private contractor. The guide provides a broad plan for protecting a wide variety of species, some of which the NPCC listed in a 2008 report:

> Species that benefit include Oregon chub, Pacific lamprey, northern red-legged frog, Taylor's checkerspot butterfly, western gray squirrel, Fender's blue butterfly, acorn woodpecker, Oregon vesper sparrow, streaked horned lark, western meadowlark, yellow-breasted chat, Vaux's swift, American beaver, river otter, Willamette daisy, checker mallow, and Kincaid's lupine, . . . cutthroat trout, Pacific lamprey, and federal Endangered Species Act–listed winter steelhead.

To protect this range of species, the TMF plan suggests different approaches to five different habitats. The first habitat is the prominent Douglas fir, which is harvested mainly for sawtimber production. Harvesting can only create openings of five acres or less to minimize impacts on water quality, soils, wildlife habitat, and aesthetics. Harvests must also leave about 5 percent of the total biomass in large clear-cuts to maintain biological legacies. Snags and fallen wood will be retained to ensure habitat for woodland species. Harvest is limited based on the rolling ten-year average, and, finally, only native Oregon species will be replanted. When the harvesting of mature Douglas fir occurs, Bowersox plans to replant with mixed conifer, including Douglas fir, cedar, white fir and Ponderosa pine.[11]

Douglas fir growth encroaches on the second habitat called out in the TMF plan, traditional oak savanna land, which was created and is maintained by burning techniques. In locations designated to be restored oak savanna, any trees competing with Oregon white oak are to be cut and any shrubs that compete with grasslands are to be removed. The creation of new oak savanna land has lead to a sizeable loss in Douglas fir through restoration logging and prescribed burning. Restoration activities are guided by a principle to create corridors to link and expand existing oak savanna habitat.

For the third habitat in the plan, oak woodlands is comprised of the same Oregon white oak as in the oak savanna but also includes Douglas fir and bigleaf maple. The only explicit work on these areas is that of thinnings that favor growth of Oregon white oak. These thinnings are exempt from the FSC limits on harvest. Removal of invasive species such as blackberry and cherry will remain a priority.

The fourth habitat, old growth forest, will be maintained for successional characteristics. These characteristics include protecting large, old trees, maintaining large snags and downed/fallen wood, promoting a diverse set of age classes and species mix. Mature trees are exempt from cutting except in cases where fire, insects, or disease pose a risk to surrounding trees. Young trees will be cut to encourage growth of large trees.

The fifth habitat at Zena is composed of wetland/riparian areas. These locations are given permission for intensive logging to protect against advancement of other biomes and are to include native planting and non-native species removal in the restoration activities. Wetland work will include stream channel revision to encourage locations for fish to maintain an active presence at Zena forest. This location is to be governed under the adaptive management techniques that encourage new practices based on evaluation of existing management techniques.

Conclusion

From the creation of the first national park to the conservation easement at Zena, preservation has been recognized as an important tool to maintain and restore the health of the planet. The NWPA influenced the Northwest's move from conservation to preservation. The BPA uses the TPL to set up some of their conservation easements to meet the NWPA, which stipulates that all land that has been destroyed by dams must be compensated by the conservation of a like amount of land with a similar ecosystem. Zena encapsulates the new paradigm of environmentalism, one for preservation into perpetuity. Zena will always be protected, and the land will never be developed. With Sarah Deumling and Willamette University, the land is in good hands.

While the BPA may only have a bureaucratic sense of place at Zena Forest, their support through financial means allows the private entities who own the land to develop a sense of place. Willamette University owes a debt of gratitude to the BPA for their financial support. The bureaucratic approach to preservation has proven to be very effective, and the program continues to expand every year. Today and expanded in the future, critical protection of endangered species and habitat will be done through actions of private individuals, funded by bureaucracy.

Notes

1. Hays, *A History of Environmental Politics since 1945*; Rothman, *The Greening of a Nation?*

2. Kieley, *A Brief History of the National Park Service*; Rothman, *The Greening of A Nation?* 101–27.

3. Northwest Power and Conservation Council, "Bonneville Power Administration, History"; Pacific Northwest Electric Power Planning and Conservation Act, 1980.

4. Karl Weist, in discussion with Kevin Bernstein; Laura Tesler, in discussion with Kevin Bernstein; Joe Bowersox, in discussion with Kevin Bernstein.

5. Willamette Basin Reservoir Study 1997; Willamette River Basin Memorandum, Attachment C, 33; Sims, *Forest Stewardship Plan*.

6. Northwest Power and Conservation Council, "Bonneville Power Administration, History."

7. Northwest Power and Conservation Council, "Bonneville Power Administration, History"; Willamette River Basin Memorandum, Attachment B, 1; Tesler, in discussion with Bernstein; Michael Pope, in discussions with Kevin Bernstein; Weist, in discussion with Bernstein; Bowersox, in discussion with Bernstein.

8. Bowersox, in discussion with Bernstein; Pope, in discussion with Bernstein.

9. "Mission & History," Trust for Public Land; Bowersox, in discussion with Bernstein.

10. Bonneville Power Administration, "Fact Sheet: Zena Conservation Easement"; Bowersox, in discussion with Bernstein; Willamette River Basin Memorandum of Agreement, Attachment G; Trust for Public Land, "Zena Forest Property"; Weist, in discussion with Bernstein.

11. Northwest Power and Conservation Council, "Council Quarterly, Winter 2008"; Sims, *Forest Stewardship Plan*.

12

Willamette University at Zena

Erica Jensen with Erik Sandersen

Scores of Willamette University community members will learn, grow, and connect to Zena through their own personal experiences. What individuals will experience cannot be known or predicted, nor can one apply his or her own connection with Zena to anyone else's relationship. To understand both the value of Zena and its relationship with Willamette University, we must examine the key players that have guided and in the future will guide Zena. These participants have differing visions for Zena. Understanding the sources of these visions will not only allow for comprehension of their unique senses of place, but will also help us get a deeper grasp on where Zena came from and where Zena is headed in the future.

This chapter maps out some of the environmental ethics underneath choices about how to build and use Zena for Willamette University. The Willamette community has a certain environmental ethos that it implements through programs such as those involving Zena and the Center for Sustainable Communities. Although most of the people involved with Zena share a broadly similar perspective on environmental issues, a variety of ethical perspectives are revealed by some decisions about what to do with Zena, from the purchase of the property to the variety of programs at the property: sustainable agriculture courses, "farm-to-fork" campus dining, restoration ecology, sustainable forestry, and more. Although Willamette University knows it has to manage the forest and acres in an economic value system, many community members strive to value Zena in different ways, differences that suggest the potential for tension in future uses of Zena.

A Fraught Purchase

Willamette acquired Zena after a lengthy and sometimes contentious process. Despite opposition from some areas of campus,

Professor Joe Bowersox and Vice President and Executive Assistant to the President Kristen Grainger persisted in fighting for the acquisition of the property. Grainger gives Bowersox much of the credit: "Joe Bowersox was integral to the debate; [he] brought people together." Among his efforts, Bowersox made a video of Zena, which encouraged viewers to "imagine yourself here." The video displayed the many possible opportunities that acquiring the property would open up. Through many months in 2008, the Zena property purchase was constantly debated and taken off and put back on the table. Bowersox recollects summarizing the situation as follows: "The train hasn't just left the station; it is way down the tracks." He urged the Willamette Board of Trustees to take action, while the Trust for Public Land (TPL) cast about for a buyer, consulting a variety of individuals and foundations; Willamette was, as Bowersox recalls, "pretty far down the list." Many months of hard work and persuasion finally paid off for Willamette, and after jumping through several hoops with TPL, the Bonneville Power Administration, and owner Sarah Deumling, Willamette University finally acquired 305 acres of Zena property for about 800,000 dollars.[1]

The purchase introduces some of the different visions and potential tensions that have and will continue to shape Zena and its uses. The moral systems that existed in the events leading up to the purchase of Zena show immense variety. The purchase of the Zena property brings together the first advocates of Zena. Their combined senses of place are discovered through their actions and bring a better understanding of the relationship they foster with Zena. For example, Bowersox's actions start to hint at a relationship between him and the forest, while resistance to the purchase suggests that not all Willamette University community members felt the same connection to Zena. Such different senses of place have continued as Willamette programs at Zena suggest different approaches to the property and different ways of connecting to it.

Pedagogy Is Paramount

Some would argue, as Kristen Grainger does, that the most important relationship Willamette has with Zena is reflected in the connection between students and the land there. "The primary tenet of Zena will be the benefits provided to the students." The Summer Institute in Sustainable Agriculture at Zena is one manifestation of this tenet. This program focuses on providing students with hands-on farming experience as well as an academic component that "examines the ecological, social, economic, and ethical implications of our food system." The program is open to any Willamette University student, introducing them to sustainable agriculture practices, plant management, composting, and other sustainable methods. Designed to introduce students to everything from the ecology of farm systems to energy conservation and farming techniques, the Summer Institute in Sustainable Agriculture program takes a holistic approach rather than relying solely on the typical classroom experience.[2]

The Summer Institute in Sustainable Agriculture gives participants a more personal sense of place: living on the property, eating

13. Summer Institute in Sustainable Agriculture students trellising tomatoes in greenhouse, 2012. *Courtesy of Jennifer Johns*

from the property, and learning from the property. This intimate relationship makes Zena seem more like a second home to students. The collective sense of place that the students gain from studying out at the Zena property is worth more than just a few school credits. The perspectives gained by students help shape the Willamette-Zena relationship as a whole. It allows students to earn a less mediated take on sustainability. This change in sense of place for each student underlies the change in the student body as a whole because it does not just promote the utilitarian benefits of Zena but also poses ideas to students that alter their mediated connection to sustainability and nature. The shift in thought can be measured in overall awareness and personal connection to Zena.

In August of 2011, a pilot program called Zena Sustainability Institute (ZSI) was launched. Not unlike the Summer Sustainable Agriculture Institute, ZSI has served as a link between Willamette University and Tokyo International University of America (TIUA). This special program gives American Studies Program (ASP) students credits that can be applied towards their education while focusing on learning about different sustainable issues in both a classroom and hands-on setting. Those involved with the process see this as both a great cultural and educational experience. According to TIUA President Gunnar Gundersen, most ASP students are from urban areas in Japan, and this program gives them the chance to challenge themselves while getting a mid-Willamette Valley American experience. "Zena is an example of the shared mission that makes Willamette University and Tokyo International University unique amongst peer universities a selling point to prospective students," says Gundersen. Without the creation of this program, ASP students' experience at Zena would have been limited to volunteer trips and some class outings because of their status as international students. The ZSI makes Zena something larger than just a part of the Willamette Valley. It has become externally valuable to many different people, something that holds its own on an international scale. The opportunities that Zena has to offer now acts as a selling point, not just to potential domestic students but also to international students, as the Zena Sustainability Institute demonstrates.[3]

Gundersen credits the ZSI program for adding to the experiences that students take back with them when they return to Japan after ten months in America. The success of ZSI after its pilot year and its popularity amongst ASP students prove that there is a niche for international students at Zena and also suggests the possibility for Gundersen and others to make beneficial changes to the program. The overall achievement that many students face at the end of their summer term at Zena is in finding the balance between appreciation and appropriation. After getting a handle on sustainability, students can hopefully change their former mindset about environmental ethics, so much so that the students take the environment out of economic measurements and understand it as included in a whole system intertwined with humans. The experiences international students take back with them will affect more than just their memories. The focus of the program will hopefully evolve to incorporating more students who concentrate on building sustainable communities.

Zena and Food

As it tends to go with farms and college students, food is often the central focus with Zena. Willamette University campus food caterer, Bon Appetit, supported Zena even before the purchase of the property. Director of Food Services Marc Marelich recalls touring the land and immediately being drawn to potential food resources already on the property such as the fig tree and the old kitchen garden. Although the relationship between Zena and Bon Appetit truly began in the fall of 2009, it is a constantly changing work in progress. As of 2012, Marelich says that less than 1 percent of the food students eat is from Zena. Challenges due to inconsistency and pricing are ever present, but those involved with this relationship have high hopes for the future.[4]

"The future of Zena can be bright, but we need to make sure we are doing a sustainable job, and work on small building blocks over time," Marelich says. From figs, plums, and apples to greens, beets, and potatoes, Bon Appetit would like to see Zena continue to diversify their crops and elongate their growing season. Although Zena is not the first local Salem area farm Bon Appetit has worked with and supported, its ties to Zena are particularly strong because the property is owned by Willamette University. Bon Appetit has not stopped with purchasing produce and other crops from Zena; it has connected Zena to other local growers, who in turn have talked strategies, given advice, and even provided the jars used for Zena honey.

If Marelich is to see the Zena he dreams of, future Bearcats could be eating strawberries, potatoes, olives, mandarin oranges, herbs, and mushrooms from the University's half-acre production field. This relationship could not be sustained and continue to grow without continued support from the Willamette University community on many levels. Peter Henry, Willamette alum and current farm manager, cannot be overlooked for his contributions to the University's farm plot. Additionally, dedicated members of the Alternative Agriculture club deserve much of the credit for maintaining and developing the property's five-acre farm plot and kitchen garden.

One of the most exciting plans that could change the lives of lucky future Bearcats is a potential internship that Marelich is working on developing with Professor Jennifer Johns. The student internship position would encompass many different steps in the food production process and would give the student an opportunity to work closely with both Marelich and Johns. Ideally, the intern would work on the Willamette University campus at Goudy Commons for the dinner and evening meals and would then work at Zena farm during the daytime several times a week. This position would differ even from that of the administrators, as the student interns would literally be involved in the whole process, from "farm to fork." Not only does this put the student in a distinctive position at Willamette University, it is also a unique opportunity in a much larger sense, not something that can happen at just any university. This internship opportunity is the latest collaboration between Bon Appetit and Zena farm, and those involved on all sides anticipate similar possibilities to follow.

Such an internship could present students with a profound change in their views towards food production, transportation and consumption. Going through all of the different processes in the internship can alter the interns' perspective in such a way that the disconnect between the finished food product and what was put into the food can be disambiguated and analyzed. This is a very pervasive disconnection that people face. If more people could see how their food gets to them, the process as a whole, a whole new world could be opened. This world could consist of healthier choices and sustainably minded decisions.

Marelich works with food, and this has shaped his sense of place at Zena. He sees it as an area to grow vegetables and fruit, to learn what works at the property and what doesn't, and to develop opportunities to better provide Willamette University students with a true farm-to-fork process. Experience, then, dictates how individuals shape their sense of place at Zena. In turn, a sense of place drives the different visions of Zena: what people want to see at Zena, programs they back, and who they would like to see involved at the property. As certain initiatives and ideas take priority over others, tension can develop. This tension stems from different individual senses of place. If the focus is directed in such a way that it only helps to develop and strengthen one individual's sense of place, then tension rightly begins to take root.

Managing an Evolving System

It is one thing to harvest Zena's agricultural bounty, but managing the 305 acres is a whole other challenge that Willamette University must take slowly and mindfully. Willamette has adopted an adaptive management plan. This approach consists of planning, then

14. Willamette student Kelsey Copes-Gerbitz taking a tree core sample. *Courtesy of Willamette University*

acting, monitoring, evaluating, and then planning again. One of the centerpieces of this plan is restoration ecology, a scientific field that seeks to restore landscapes to a past, often pre-industrial, state. The goals of the restoration ecology at Zena are simple: renew and restore oak savanna habitat. Professor Karen Arabas has a special connection to both oak savanna and restoration ecology:

> The idea that landscapes that we encounter when we are very young stick with us in unexpected ways. For me, the open savanna landscape was a part of my childhood and created a strong sense of place in me. When I came to Oregon in 1996 I felt "at home" in the valley because it was so similar to the cherished landscapes of my childhood. So from a purely emotional standpoint I have a connection to the oak restoration project at Zena. But I would caution that it's more complex that just a visceral response to the landscape that drives my interest in restoration—I also draw on the ecological knowledge that guides our restoration efforts.

Arabas clearly feels strongly about the value of restoration ecology, and those values stem from a variety of roots, from childhood experiences to professional interests and training.[5]

But there are other ways of approaching the management of the Zena property, including sustainable forestry of Douglas fir. Sustainable forestry is a system of strategic planting and thinning that abides by the decline of the oak savanna habitat and maintains Douglas fir in the area. Although restoration ecology and sustainable forestry share a broad concern with the health of the environment, they also suggest some differences in values. One of the benefits of sustainable forestry is the potential economic gain from selling the lumber gained from thinning the Douglas fir. Restoration ecology, on the other hand, provides less economic benefit, at least in the short term; its clearest immediate value comes in the form of education. The two different strategies presented are both possibilities for Zena's future, and we can see two different senses of place that, if not recognized and reconciled, might lead to tension.

Post-Purchase Zena

Since the purchase, Zena has morphed into an essential part of Willamette University, providing professors, students, and community members with an ecologically rich property and a wealth of potential opportunities. The different visions shaped by the acquisition of this property have created many beneficial programs and opportunities, and at times, these visions may overlap and possibly even conflict. To some, Zena is a place where knowledgeable and experienced farmers, foresters, and scientists can practice and develop their research and skills, while to others, Zena may be where people are introduced to sustainable agriculture and the environment and development of related skills. It may also be a place to experiment: with programs, with food, and with different sustainable and organic techniques. Zena can be and is all of these things. The different visions held by different people are shaped by what they dream of seeing at Zena, as well as their personal and professional backgrounds. In short, different people have brought and developed different senses of place to Zena, and those different ways of interacting and connecting with the property will continue to play an important role in how Zena is incorporated into the Willamette community.

Notes

1. Kristen Grainger, in discussion with Erik Sandersen, October 2012; Joe Bowersox, in discussion with Erica Jensen, March 16, 2012.

2. Grainger, in discussion with Sandersen; "Summer Institute in Sustainable Agriculture," Center for Sustainable Communities.

3. Gunnar Gunderson, in discussion with Erica Jensen, March 8, 2012.

4. Marc Marelich, in discussion with Erica Jensen, March 16 2012, and in discussion with Erik Sandersen, October 2012.

5. Karen Arabas, in discussion with Erik Sandersen, November 2012.

13

Personal Stories of Zena

Elise McGlone and Lauren Vermilion

Tucked into a valley where twin peaks peer at the projects and classes from Willamette University, Zena is part of a larger patchwork of red barns, green grasses, plump purple grapes to pick, trees that tackle hillsides, and horses clicking their hooves. As Zena becomes more distinguished, it impacts the surrounding community by allowing them to foster their own plans for its future purpose. Different social groups, ethnicities, income levels, and general life experiences inform the views these individuals have formed about Zena: a Grand Ronde tribe member who has committed himself to soil composed of his ancestor's bones, a man who immigrated from Mexico, a representative from the United Methodist Church of Salem, a business owner dedicated to sustainability, and a recent daughter of this land: all are connected through Willamette's Zena. Their individual positions within these landscapes lead them to perceive the same physical space in different ways.

This chapter examines outsiders' evaluations of Zena and their assessments of Zena's potential. These ideas could be implemented by Willamette, and thus Zena could better serve the community at large. This spectrum of voices divulge their life stories within the landscape, how it has shaped them, their notion of what purpose it serves, and what they believe the best course for Zena could be. These oral histories can be interwoven and analyzed in order to decipher larger patterns of the community's desires. Although each interviewee had a different notion of what Zena is, all agreed it was positive for Zena to be owned by Willamette. Regardless of social status, all interviewees had a sense of stewardship towards the valley—the variance was in the way that this was implemented. A tool used for analysis is "sense of place," which, broadly defined, is the evoked emotional or intellectual response to a certain place or entity within the landscape. This chapter makes two contributions to our understanding of Zena and sense of place. First, this chapter shows how a person's life experiences, especially employment, shape that person's sense of place. Second, this chapter suggests how socio-economic values are reflected and reinforced by the landscape. By examining this dynamic between people's backgrounds, their sense of place, and their visions for Zena, this chapter offers insight into surrounding community's perspective on Willamette's Zena: what it is, what it can be, and what it should be.

David Lewis

A dual-citizen of the sovereign nation the Confederated Tribes of Grand Ronde and the United States, anthropologist, writer, museum curator and educator David Lewis, Ph. D., has a unique insight on the Grand Ronde. He currently serves as the coordinator for the Cultural Resource Center. Although he graduated from the University of Oregon with a Ph.D. in anthropology, many of his fellow tribes people lack that same level of education. Part of Lewis's mission is to inspire and to inform about Native Americans across all social boundaries. He has published multiple books and essays on Oregonian tribes. He works as a valuable ambassador for his people, retrieving artifacts from different US government institutions and restoring them to their rightful place. Utilizing some of these recovered artifacts he is able to reinstall old traditions, and he plants conventional crops like camas for harvest. As he explains, "I work with various organizations trying to restore a traditional landscape. There are already efforts in Corvallis, and Wilsonville, Portland and Eugene, Ashland." Through these efforts Lewis contributes to these larger communities as well by restoring cultural values as remembered in the recent past. He works in the tribal context of service; he has dedicated his life's work to it. He helps schools, governments, and other tribes appreciate and understand tribal history and culture.[1]

David Lewis, like other members of the Confederated Tribes of Grand Ronde, feels a strong connection to place through his ancestry. As he explains, "Perhaps thousands of generations have occurred in the last 14,000 years. And that means almost every piece of dirt you see here has some relation to me. Because when people are buried, the body doesn't just disappear. It's all processed by the environment and turns into dirt." As Lewis observes, the Kalapuyans have a strong historical connection to the land of the Eola Hills. "[The Kalapuyans] didn't want to be moved to any other place but this place; this is their place." This physical space was worth more to the Kalapuyan people than their lives. The land bonds them by giving them a physical space, as Lewis explains. "Knowing that at some point in the past Americans tried to wipe us all out several times, the only way we can maintain that culture resulted in spiritual connection with the land to be able to thrive as a people, as a nation of people." Physical boundaries of the land create a tangible space for hope, inspiration, and connection as a community.

Their connection to the land is demonstrated through their perception of what should be done with Zena. David Lewis explained that without a physical space to connect to one another and to practice their traditions, their tribe disintegrates. Their culture is embedded in the landscape and in oral histories. As far as Zena is concerned, Lewis's position is that Native Americans should be present in the discussion and dialogue to restore the oak savanna habitat. "The problem comes in when organizations take it upon themselves to decide what the natural environment is without any interaction with the tribes. . . . You have to have that cultural connection in order to truly create this sort of this pristine landscape." Furthermore, he talked about the type of decisions that the tribe would have to make when they burned and why they had chosen to alter the land in this way. This ecological knowledge is tied into their culture, and that resource could be utilized by Willamette. "We were making a choice for what species to keep and what species to get rid of or renew and in a cultural way, every ten years so or every year perhaps in some places. It will be impossible for anybody, any organization to restore land without native involvement." Lewis has the ecological resources to reestablish the native landscape, and Willamette University may have the scientific methods to recreate the landscape. Lewis elaborated on how other restoration projects work and how it could benefit the Grand Ronde tribes and Willamette symbiotically by bringing native culture back to the Eola Hills along with the vegetation. This would include helping to harvest certain crops and some traditional ceremonies. "Without cultural context of each species that exists at Zena, the property is just a memorial instead of a living landscape." Lewis would prefer to see a landscape that is used by the native people, which would also benefit Willamette, as it would enable important cultural work to be enacted while maintaining the landscape. In this way the land would be mutually beneficial and utilized sustainably.

Bill Olsen

Bill Olsen showed up at Willamette's Bistro coffee house ready to tell a story. He alluded to his past as a government employee with a warning that I might need a filibuster to stop this oral narrative. His voice the vehicle, I the passenger, he started driving down the I-5 into sunny Orange County, California at the height of the environmentalist movement of the 1970s. "I grew up in the suburbs. . . . I was in planning, and most in planning have some interest in conservation. I think that you get caught up in a general concern that all the land is open to development. What's the Joni Mitchell song? 'They paved paradise, to put up a parking lot.'" His work sought to confront this problem. "I headed a group called the Agriculture Preservation Task Force. . . . There was concern about all the agricultural lands being urbanized and preserving open space and agricultural values." During his time there he discovered that the best way to preserve land was through conservation easements. Land was too vulnerable if it was left to the farmers to defend all on their own. After retirement, Olsen left the sunshine in order to be greeted by bright smiles of grandchildren in Salem, Oregon, where he now volunteers for the First Methodist Church of Salem. Olsen organizes presentations for the Sunday School. A few years ago, one such presentation was given about Zena. Although he has seen Zena while on various wine tours, he has never walked on the property itself.[2]

When questioned on where he feels most connected to God he replied, "Things that create awe. I am relatively small and insignificant in the grand scheme of things. . . . This will outlast me and everybody else. I think it inspires me." Destroying land sends a message to God that we vainly attribute our own value system onto the variety of creatures he created. He referenced the Environmental Mandate of the United Methodist Church. It begins, "All creation is under the authority of God and all creation is interdependent. Our covenant with God requires us to be Stewards, protectors and defenders of all creation. The use of natural resources is a universal concern and responsibility of all as reflected in Psalm 24:1: 'The Earth is Lord's and the fullness thereof.'" He sees nature as a therapist; it is a redeemer, a safe place. This in conjunction with his past as an urban planner creates his philosophy that land must be utilized in order to produce the necessary goods to support humanity, but we must also protect biodiversity and open spaces. He believes that land should be used but not developed too rapidly; it should be thoughtfully utilized.

To Olson, Willamette's Zena is a preservation ground, serving the entire landscape by providing a pristine landscape that is not threatened by development, taking pressure off of other farmers and their families to be the source for preserved lands. "[Farmers] didn't want to be left holding the bag if their kids didn't want the farm. That's why I think purchasing something is best way to preserve and the best way for the community. And that way we aren't saying to someone: be our agricultural bank for us." He urges Willamette to extend the influence of Zena to the surrounding community so that they can use this land as well to enrich their lives. Since Zena thus becomes the "agricultural bank" for the community, Olsen's affirmed the restoration projects at Zena. "To the extent that you are setting aside a relatively small percentage of the land, it is a good idea. On the other hand if you were saying everyone in the state has to do it, then you would be forcing your ideas onto others, which is detrimental." However, if it was a larger portion of land, it should be utilized for large-scale farming in order to provide for agricultural needs locally. "I would see it as a resource for students like me brought up in the city to connect to the nature. Even now I'm not aware of the things I grab at the supermarket shelf. I think students should start learning how to live locally." Zena is a resource as a preserved green space, and it should serve as a way to inform students about agriculture and local produce. His stance is to conserve land for agricultural use and protect the biodiversity of local native species.

Kari Ramey

Kari Ramey came in contact with Zena through what can only be described as destiny. She is a Seattle-raised marketing consultant and interior designer who always dreamed of living on a vineyard, of waking up to the sun casting her red blush across buds wet with dew, water droplet mirrors that make the green leaves glow, and grapes that tell stories of the soil through flavors found within. Kari's husband Tim worked on Wall Street in New York while, Kari says, "We spent a tremendous amount of time in event centers. Saw different wineries that did weddings and auctions. I also spent 10 years in the marketing business in the Junior League doing non-profit events on the East Coast." At a dinner party five years ago, a phenomenal bottle of wine intoxicated them, and they learned that the vineyard was for sale. "We thought to ourselves, 'Hey, we can go to Oregon for the weekend; that will be fun.' We flew out here on a bit of a whim. We met Pat O'Connor, toured the farm and by the end of the weekend, we were like, 'Yeah I guess we're gonna buy it.'"

They started out as "gentlemen farmers," but even with two salary-earners working full time they could not keep up with the bills. "The worst part is the financial issues. We want to gain equity, and we should be saving for retirement, and we still have a daughter to put through college. We just started health care for all our workers.... A couple weeks later the insurance company raised the premium 20 percent. We aren't taking it away, but it's still like 'ouch.'" This awareness of their financial dependency on the land inspired Ramey's "Wedding & Events Center" enterprise, which Ramey sees as a benefit to the larger mid-Willamette Valley area. "My business alone sold 350 rooms at the Grand Hotel, plus a couple million dollars for catering, and the bus drives, the florists, the entertainment. If I do 70 events a year imagine how many jobs that is.... One third of my business is out of the state, people who would never come to this region.... It's a pretty dang good business." Kari has managed to secure a beautiful vineyard in the middle of Salem where she can play in the dirt and ride horses too.[3]

Ramey believes fruit for outstanding wines can only come from beautiful origins. "When Tim and I first moved in there were a lot of chemicals. We took them all out, brought the liveliness back to the soil." Ramey sees herself as a caretaker. "We are charged with delivering this parcel of land to the next generation with healthy soils, intact biology, and a vineyard that has a brighter future ahead than even its greatness of the past." In everything they do, they approach it with the attitude of: "We're doing this forever. Don't cut corners, and don't make short-run decisions." Kari is clearly enamored with the land she lives in, and furthermore wants to ensure that it is there for future generations to feel the same. She describes it as her cradle, and says that it's invigorating and indispensable. "I get a sense of purpose, a sense of coziness cradled in the Eola Hills.

I have no plans to leave ever, and there will be some stone with my name on it." Zenith Vineyards protects Kari in turn, and gives her great joy.

Ramey's view on Willamette's Zena combines her economic sense with her emotional sense of place. In her opinion, Willamette needs a multi-layer business, and it needs to think about how to get others to pay to keep Zena afloat. The property should make money and be economically sustainable, perhaps with cooking classes, tourism, or art. "Knock that house down; make a studio somewhere to take the trustees. Have an open classroom, a decent kitchen so you can cook." This perspective derives from a deep passion for balance. "Economically sustain the agriculture. We strive for the complement; never negate it. . . . If you can't economically support it, it can be taken from you. With the WU property I am hoping that you will be building a sustainable visiting type of economy. . . . Find a way to make it pay." Ramey also has an invitation for Willamette: "When you do this farm research, always do it with a mission. You have the privilege of researching, but don't be frivolous; think, 'How can we help our neighbor?'" We are part of their web.

Pedro Martinez

Twenty years ago when Pedro Martinez was working in one of the poorest states in Mexico, his brother and sister migrated to find work in the United States. A letter came and blotted in black ink Pedro's name, beckoning him with the incentive of the opportunities in Oregon to make money and reunite with his family. After traversing through various states, he eventually arrived in the Willamette Valley. Eighteen years have passed since Pedro Martinez started working at the O'Connor Winery, and although the title shifted to Zenith Vineyard, his roots on this plot of land were planted. Kari Ramey made the decision to promote him to farm manager because of his tenure and his obvious care for the land. He is a lifelong gardener; instead of a wedding ring, the mark he bears is a green thumb. Of his work, Martinez says,

> Me gusta mirar las plantas, crecer, a grandes de chiquitos.
>
> I like to watch the plants growing to grand things from these little guys.[4]

This became apparent when questioned why he chose to work as a farmhand instead of searching for a more lucrative field:

> Me siento bien cuando trabajo con las plantas en la naturaleza. Me gusta cuidarse a las plantas, porque al final yo sé que las hizo. Para construir una casa es algo diferente. Estoy feliz aquí en el aire libre. Quiero trabajar aquí. Siento muy afortunado.
>
> I feel good when I work with the plants out in nature. I like to be the caretaker for the plants, because after they have finished growing I know that I have cultivated them. To build a house is something different. I am happy when I am in fresh air, and I want to work here. I feel very lucky.

Martinez enjoyed working on this farm more than others, because he agreed with their sustainability practices. His main concern is to keep these plants alive because he feels a great sense of pride towards the plants that he is able watch grow and, after they bloom, know he played a vital role in their creation. His reasoning for not using chemicals was shaped by the way the plants appeared, as he explained:

> Creo que es mejor para las plantas, es mejor para la gente, para todos. Las plantas están mejores sin las químicas. Creo que nosotros debemos proteger a la tierra. Me encanta su tierra y quiero que estaría aquí por muchos años después de mi vida.
>
> I think it is better for the plants and humans as well, for all of us. The plants grow better without chemicals. I think that we should protect the Earth. I love this land and I hope that it will remain for many years after I pass.

His commitment to the land on which he works is driven by a love for the cultivation of plants.

A road and some shrubbery were the only things that barred our eyesight from Zena, and I pointed to it as I probed about his impression of Zena under the direction of Willamette University, for he was here before the conservation easement and witnessed any changes from his placement at the former O'Conner Winery. I asked him if he thought that there was justice in the fact that while he worked for money, we worked simply to learn. Valuing the education gathered at Zena, Martinez replied with eloquence,

> Es bonito para aprender muchas cosas. Hacer las cosas está bien. Tengo como dieciocho y todavía quiero conocer las más de las plantas.

It's beautiful to learn. It is good to do these things. I have worked here for 18 years and I still want to know more about the plants.

A sense of balance of responsibility reigned in his response. Martinez believes it is viable to use the land for studying if the information is shared with the entire community. With this knowledge he will be able to better grow the plants he loves.

No quiero investigar, pero quiero saber más. . . . Creo que es bueno para otra gente aprender y compartir la información con todas las haciendas.

I don't want to do the research, but I want to know more. . . . I think that it is good for others to learn and share the information with all of the farms.

In this way, Willamette's research is actually serving him, because his role is to cultivate plants, and acquiring new information about the Willamette Valley and sustainability practices allows him to perfect his farming methods. If both parties contribute in their specialized field, the combination of these factors will better serve the Willamette Valley as a whole. Knowledge in this way is not just bound in books found on our shelves; Martinez professes his superior knowledge of the landscape and farming. His opinion is that a combination of multiple types of education is better. However, Martinez did not share Willamette's enthusiasm for making choices between different species.

No está bien para destruyen a los árboles . . . ¿Por qué escoger algunos tipos de los árboles?, y también ¿por qué destruir algo?, todavía es una planta.

It's not okay to cut down the trees. . . . Why choose some types of trees over another, and furthermore, why would we destroy something? A plant is still a plant.

While environmentalists may claim there are specific reasons to destroy certain types of trees, this does not resonate with Martinez, because he values all life and all food sources, and doesn't agree with wasting anything. Environmental science students, on the other hand, have been taught about native species and keystone species, and we are taught to value certain species more than others. There is, in short, a cultural gap, and because of that, Willamette's actions appear very contradictory to conservation, as suggested by

Martinez's description of his understanding of the burning practices and the restoration of oak savanna.

> Es difícil, porque es como, por quitarte la mano aquí para ponerte otra mano que no es su mano. No, no te va a gustar en los años que no está bien tu mano y todo está la misma. Dejarte el original es el mejor.

> It's difficult, because it's like, you are cutting off your hand to put on a new hand that is not your own. No, you won't like it in a few years that your hand is no good, and it's all the same. Keeping the original is best.

He further explained this traditional phrase, "cutting off your hand," which is analogous to "shooting yourself in the foot": by removing and replacing the species that already exist, we are really hurting ourselves and the land. It is not our place to continue to keep choosing what plants should exist and which should not, because historically such decisions have not always been right. Knowledge is fluid and oftentimes the scientifically known facts become disputed and all subsequent actions must be reversed. These patterns of decisions of destruction and reconstruction help explain the current emphasis on reduction in Douglas fir stands at Zena. Yet we still perpetuate the notion that we know what is best for the land. Instead of taking away the original, we should allow for nature to choose what happens, because "the original is the best."

Anne Walton

Down a gravel road just south of Willamette's Zena, hundreds of bright yellow daffodils spring up and cast the spell of sweet fragrance from the soil that lines Anne Walton's house, ready to spill the secret of what lies beneath the surface before she can reveal them to us. The past generations of farmers live on in the present with the presence of these same flowers, which have bloomed for generations. Each spring Walton carries a small bouquet of these daffodils to the cemetery and honors the graves of former Zena residents. Much like the flowers, those who have lived at Zena will inevitably return to this sacred space. Anne Walton's family first arrived at Zena in November 1902. Their farm was self-sufficient and this working property remained their primary home until 1918, at which point it became a rental house. Walton's family, meanwhile,

15. **Daffodils at Anne Walton's property.** *Courtesy of Elise McGlone*

moved to Roseburg, Oregon. Farming was a "family value" instilled in her from as young as she can remember. "It's a learned thing; it really is values of your parents that say, it's fun to see that bug get bigger." As a small child, Anne recalls, she would follow her father and grandparents around the yard to peer down at the spectrum of plants and animals. Visits to see her grandparents during her youth planted a seed that would grow into her love for biology and forestry. Inevitably, it inspired her to migrate back and purchase the property for her own family in 1998. She reiterated, "It's my background and my love of biology, and my love of riding horses and forestry." Anne's personal history is written in the hills.[5]

When she goes out on her property, Walton is not just connecting with the land but also with a place that was her family's livelihood for decades. "It's a smell of dried grass and kind of ripening apples—you know, all that ethylene coming off. There's a particular smell. . . . You could blindfold me and I would know I was there." Every day, this property's history impacts her interactions with the land: "It's truly a privilege to be on the land that your people were on. It's very hard to describe; it gives it a value and a sense of people that are gone now that have lived there. It's just indescribable, and it dictates your choices and decisions that you make." Walton remembers her grandfather telling her when she was ten years old, "Look around. I want you to love this place as much as I do." This background of her family—the feeling that she was meant to be there—informs her sense of place that the area that is Zena is her home. It connects her love for her family with a physical space.

Walton decided to share her space with students ten years before Zena was purchased. "You have that farm there [referring to Willamette's small farm plot] because when I bought it in 1998, . . . I immediately came down and said, 'I own this piece of property and if you can come down and use it as a field station go right ahead.'" Having science classes come out, explore, and do tests on her land furthers Walton's values about the importance of nature as well as of teaching the next generation about the magic of Zena. In her opinion, working with the land is one of the most valuable parts of an education at a liberal arts college such as Willamette:

> Everything you learn [at Willamette] is to appreciate those things in the world, and so it fits in perfectly [with the curriculum], because what's more natural than how you'll care for yourself or understand your food? It doesn't matter what you're going to do to earn a living, whether you play a cello, it's really important.

Walton sees a vital role for Zena in a liberal arts education. It is as interdisciplinary as imagination can be applied. Walton believes that Zena should be used to raise awareness in the student body of the natural environment and their role in it through everyday life and food consumption. Her family owned this farm for generations, and she attended Willamette University. Her sense of place is that Zena is home; her viewpoint is that Zena is a forest, a farmland, a classroom, and her backyard. The farm and forest should be used for research because Zena is so universal and can bring collaboration between students.

Zena's Zenith

Zena serves the community as a keystone among these emerald gems of treetops, the quilt of quintessential crops, the army of fortified old farm houses, the testaments of tenants. It is a collaboration, a conglomerate, an entirely new creation. Its potential is only limited by the bounds that we arbitrarily assign. By looking at how the surrounding community has historically been impacted by the Eola Hills and continues to be influenced, it is possible to glean new meaning. The environment, this plot of land, is a tangible way to adhere to the Willamette motto, "Not unto ourselves are we born," by listening and addressing the voices of the community around us.

Notes
1. David Lewis, in discussion with Lauren Vermilion, 17 October 2012.
2. Bill Olson, in discussion with Lauren Vermilion, 2 October 2012.
3. Kari Ramey, in discussion with Lauren Vermilion, 3 October 2012.
4. Pedro Martinez, in discussion with Lauren Vermilion, 7 November 2012.
5. Anne Walton, in discussion with Elise McGlone, 3 March 2012.

14

Stories of Zena from the Willamette Community

Heather Smith with Elise McGlone

Willamette University's commitment to its students, staff and faculty, as well as members of the surrounding community, is built on the foundation of its motto—"Not unto ourselves alone are we born"—and its values:

- the dignity and worth of all individuals
- a commitment to diversity, service, leadership, and sustainability in communities and professions
- the ethical and spiritual dimension of education
- education as a lifelong process of discovery, delight, and growth, the hallmark of a humane life

These standards blanket the entire university, and in recent years they have been applied to encompass the environment through a commitment to sustainability.[1]

Just outside of the city limits lies another wide sphere of opportunities: the Willamette Valley and, nestled within the Eola Hills, Zena Farm and Forest. A key player in the university's commitment to academia and the environment, Zena is a new character in Willamette University's story. It is a "sustainability laboratory," a place where people of different backgrounds can come together to explore their role in promoting a healthy environment. Although the landscape of Zena shapes this exploratory process, the individual experiences of Willamette community members are influencing both the physical and educational future of this unique space. An individual's role within the Willamette community shapes his or her perspective of and connection to Zena, creating a different sense of place unique to how they understand Zena as a part of the WU community at large.[2]

Anthropologists, historians and other scholars suggest stories as a new method of examining the past, one grounded in place. Keith Basso suggests that through narratives, individuals are able to explore the specific details and physical aspects that link them to a particular place, a particular time. As the place changes, so will the narrative, providing a point of reference and a method of looking into the events and circumstances that led to these changes. The feelings and emotions established through experiences are embedded in a sense of place, and with every passing experience our sense of place evolves. In this way, William Lang explains, as sense of place influences experience, experience also influences sense of place, creating a reciprocal relationship. Lang, however, also recognizes the limits of historical narratives, because the focus of each story is channeled through a single individual's perception. If only one voice is heard, the scope of a place becomes reduced.[3]

This chapter provides a spectrum of stories about Zena told by a variety of Willamette community members: the University president, a recent graduate and current Zena manager, and professors from a range of disciplines. The stories told in this chapter aim to widen the image of how Zena fits into the Willamette community at large. Each story voiced here hopes to provide the reader with a better understanding of the many different ways people form a sense of place at Zena. The stories of these individuals come from their experiences as members of the university and from their lives preceding Willamette. Although the individuals featured in this chapter all desire for Zena to develop in accord with the mottos and values of Willamette University, their interpretations of how this should come about vary. Within this variance lies the potential for tension. If stories are censored, left untold or unheard, this potential for tension will only increase.

President Stephen Thorsett

In 2011 Willamette University welcomed its 25th president, Stephen Thorsett. President Thorsett grew up in Salem and, as the son of Grant Thorsett, a past professor of biology at Willamette, he is no stranger to the standards and values of this university. Gazing out of his fourth floor office window towards the golden explorer atop the Oregon State Capitol across the street, President Thorsett recalls his first visit to Zena. It was summer, July 2011, only a few weeks after his arrival to Willamette. The faculty running Zena were working with the state to complete a controlled burn, hoping to eradicate invasive species and promote the ecological restoration of habitats like the oak savanna. Since then he has returned to Zena two or three more times and each time is exposed to a new aspect of the place and what it has to offer.

Although his connection to the actual physical space of Zena has been limited, President Thorsett believes strongly in the opportunities that Zena presents to both faculty and students of Willamette University. Zena, he explains, provides a unified space where people from different backgrounds can come together to discuss and explore sustainability through varying perspectives. The interdisciplinary use of Zena is reflected, through President Thorsett's eyes, in the growing number of students who travel to the farm and forest each passing year. "Approximately 300 students went to the farm in 2010 and then 400 in 2011, and this past summer [2012], 350 students went to the farm", he explains. "This is a place where students with lots of different interests come together." The way in which Zena brings people together from different programs, both within the Willamette population as well as from the surrounding community, is, as President Thorsett describes, "pretty unique for a university in the West."[4]

For President Thorsett the benefits of Zena are both intangible and tangible, coming not only in the form of education but also in resources. President Thorsett highlighted the telescope located at Zena, saying, "You can think of the dark sky as a resource that is being lost. Having that space were we can reclaim relatively dark skies near Salem is really important." The deep passion that President Thorsett has for the night sky reflects his extensive background in astronomy, and he recognizes that as it is slowly disappearing, the darkness found over the fields of Zena is priceless. President Thorsett knows that there will be many different visions for Zena and describes its future as somewhat of a "blank canvas" that must be filled by the faculty and students of Willamette. During its brief history, "it has been used in ways that nobody would have imagined during the proposal to acquire the property," and President Thorsett looks forward to the future of Zena.

Since his return to the school, President Thorsett's connection to Zena had been limited; however, he realizes the deep value of having a space that is unique to a school like Willamette. The opportunities and resources it provides are irreplaceable. He understands the importance of experience and professes that the best and only true way to understand Zena is to get in a car, drive to the farm and talk a walk. "Ideally," he says, "[I'd explain Zena] by getting someone out there during different seasons and different times of year, get them out there at night when it's quiet so they can see the stars." It is a place that must be touched, felt, and experienced.

Silence, darkness and shining stars in a clear open sky: these are just some of the resources that President Thorsett recognizes as indispensable aspects of his sense of place at Zena. He understands that in order to maintain these types of resources we have to manage the land, and through this need for management, the creativity of students and professors can grow.

> It has always been a managed landscape, and thinking about how you approach a managed landscape like that—where there's not a right answer as to how it should be managed, [and] you get to answer all the policy questions about how you balance use of the land and how you integrate the human aspects of the land with the natural aspects of the land—makes it a really interesting site.

His sense of place at Zena is shaped by his position as the President of Willamette but also through the experiences he has within his expertise in nature and science.

Peter Henry

Peter Henry, a Willamette graduate with a degree in biology, was first introduced to Zena as a sophomore, just a year after Willamette purchased the farm. As a lover of nature and science Peter quickly became involved in Zena, first through a project that worked to clean the invasive species found on the property, and later as the founder of Willamette's Compost Club, which took food waste from the main dining hall, Goudy Commons, and brought it to a pile on the farm. Today, a year after his graduation in 2011, he is the farm's manager.

Peter finds many personal benefits from his connection to Zena. "It provided me with a job right out of college," he says, "but more then that, at Zena I am forced to know everything that we are doing. I need to understand the science of agriculture so I can explain it to people I am working with, as opposed to working on another farm where I probably would not have been forced to have the same depth of knowledge." Peter enjoys working in an atmosphere where his colleagues, Willamette students and professors, do not buy into what he calls the "organic religion," in which organic is understood as inherently better. Spending time in the soil and among the plants helps Peter gain a greater understanding of one's place within nature. Peter explains, "By working on a farm you are playing a part in belonging to this Earth, as opposed to taking a hike where you walk through the woods to appreciate the aesthetic beauty; in growing your own food you are actually participating in your own survival."[5]

For Peter the many benefits of Zena, while not currently used by all Willamette students, are definitely available to them if desired. "There's a lot of room for growth," he says, "and [because of Zena's short history at the university] the message that anyone can get involved is still getting out to the Willamette public." Discussion of the environmental issues surrounding our food industry are becoming more and more prevalent on campus and, for Peter, Zena fits into the goals of Willamette as a liberal arts institution by providing another avenue to understand the world around us.

Peter spends most of his time at Zena in the greenhouses and on the patch of agriculture slopping down an open field, divided by rows of vegetables and partially shaded by tall sunflowers. It is evident that his experiences are deeply influenced by the agriculture program at Zena. Soil, plants and a connection to the Earth allow him to feel as if he is part of Zena, participating in a process that not only benefits the land but also those who take from it. As a participant in that process, he feels he does his part to nurture understanding about how the relationship between the non-human natural world and humans can aim to be more mutually beneficial. Through experimenting with sustainability and organic agriculture he is setting roots at Zena by giving back to the land and creating a platform on which others will be able to build.

Professor Matthew Nelson

Matthew Nelson earned his MFA from the University of Utah and is currently a visiting professor of choreography and dance at Willamette University. In 2011 he received a grant from the Willamette Center for Sustainable Communities to create a project that promoted new forms and expressions of sustainability on campus. In the summer of 2012 he and a group of selected Willamette students created the "Permaculture Dance Project," which as he explains, "[worked] to investigate the principles of permaculture, a model for sustainable living, through movement." In the video Professor Nelson defines permaculture as "a branch of ecological design and engineering which develops human settlements and self-maintained agriculture systems modeled from natural ecosystem." Professor Nelson believes that the principle ideas from permaculture correlate strongly with many of the current activities at Zena and should be considered as guidelines for all future actions.[6]

Sitting outside the Willamette Playhouse, the building of theater and dance, Professor Nelson reflected on his experience at Zena. "Movement is place based," he states, "the environment changes everything in dance." For this reason, Zena Farm and Forest was an ideal location for Professor Nelson to conduct his project. He choreographed the movements first and then had to work within the space of Zena to assign the principles of sustainability to each movement. The landscape he describes had great impact on these assignments.

> The first principle that I investigated was *observe and interact*, and it's the idea that you need to see a place before you begin making changes to it or choices within it. So the first day we walked around and we found this field where we started working. It was a spot with a beautiful view of the hills and where the grass wasn't so long that we were lost in it, and also where we felt like we weren't doing any damage that would cause a problem.

Grasses, vegetables, and the slope of the land—these were all things considered by Professor Nelson when he chose locations to dance. The land played its part in creating boundaries. The poison ivy and blackberry brambles prevented Professor Nelson and his students from exploring certain areas of Zena. Environmental elements such as these are a reminder of how, just as much as we influence the land around us, it influences us, constricting us and shaping how we understand it.

For Professor Nelson, the dance project is part of the bigger picture of what Willamette is trying to do with Zena. He suggests, "We are seeking to understand what sustainability is about and how we begin to teach it and embody it." His desire for Zena is to see it as a full permaculture site. In this vision students would be able to spend more time at Zena with a possible residency program. More in-depth exposure to a site like this would allow students to investigate nature through the "tamed" landscape that is Zena and "experience what it is to create a living ecosystem."

Movement defines Professor Nelson's sense of place at Zena. His background in dance has given him the tools to look at a landscape and realize the implications of human action and movement within the space. From blackberry bramble to tall grasses, respect should be shown for nature's physical boundaries. In some cases we can transform these boundaries to allow movement, but these transformations, whether they simply harm small plants or drastically alter ecosystems, have consequences. In many ways these consequences are inevitable. However, just as the dancers involved in the project were aware of the impact of their presence, people at Zena must be aware of the consequences of their actions. Professor Nelson understands the physical space defined by man-made boundaries as a place for experiment and growth in our understanding of sustainability and our role as stewards of the land.

Professor Rebecca Dobkins

Rebecca Dobkins has a long background in Native American society and culture. Before coming to Willamette University in 1996, she was a researcher at the Smithsonian Institution's National Museum of Natural History, and now continues to investigate the culture of different indigenous peoples, human rights, and the environment as a professor of anthropology. Although Professor Dobkins does not have a strong connection to the physical space of Zena, she has a close relationship to the Confederated Tribes of Grand Ronde. Her understanding of how Zena fits into the Willamette community has been greatly shaped through her close relationship with these tribes and their perceptions of how Zena came to be part of the university. And as Professor Dobkins said to me, "Perception is important; it is all about perception."[7]

Professor Dobkins, through her research and conversations with tribe members, believes Willamette's motto corresponds with the values of native people. "There is an understanding [within native culture] that you are not born unto yourself; you are born into generations of people who come before you—your ancestors—and, you are born for the generations that come after you," she explains. Her desire for Zena's future is for Willamette's community to take the meaning of our motto and genuinely consider all the generations of people who for thousands of years lived in the Willamette Valley before us. This would mean establishing a relationship with the Confederated Tribes of Grand Ronde and asking whether they would like to be involved in or represented at the Zena property. This relationship, Dobkins believes, would only enhance the interdisciplinary goals of Zena by creating a space for cultural education and a more diverse interpretation of how to interact with the landscape is a sustainable manner. There are many beautiful aspects of Kalapuyan culture and tradition that, if applied to Zena, would only enhance the physical space and the experiences people had there. From a blessing ceremony to the use of Kalapuyan names on the property, Professor Dobkins hypothesized many ways in which representation of native culture could be incorporated at Zena. None of them are possible, however, without establishing a meaningful relationship.

Professor Dobkins's narrative makes one consider the role of history, background, and social title in developing a sense of place. Her deep-founded respect for and knowledge of the native tribes of the Willamette Valley has influenced her sense of place by connecting it to people whose ancestors lived on the land, and the sense of place they have from this long connection and history. In the same manner that we must bring together Willamette stories of Zena, we must also consider the stories of those before us. Stories that come from members of the Confederated Tribes of Grand Ronde are equally important as those voiced by members of the Willamette community, and just as they have influenced Professor Dobkins' sense of place a Zena, they have the ability to add many patches to the quilt, creating a bigger picture for the future of Zena.

Professor Joe Bowersox

Joe Bowersox is a professor of politics and environmental and earth sciences at Willamette University. Echoed by the dirty spades on his office floor during our interview, he is an advocate of hands-on learning and a commitment to the environment. He is heavily involved in Willamette's commitment to sustainability and serves on the Board for the Center for Sustainable Communities. Professor Bowersox was the individual at Willamette University who was initially approached about Zena, and he helped organize the campaign to purchase the property in 2008. Since the purchase, he has continued to be heavily involved in Zena and currently runs the forest management plan. From administrator to "gravel-hauling gofer," Professor Bowersox says he will fill whatever shoes necessary to help Zena succeed.[8]

Professor Bowersox works alongside other Willamette faculty, like Professor Karen Arabas, to restore at-risk species and their habitats. Ecological restoration projects at Zena include the few remaining oak savannas found on the property as well as Fender's blue butterfly, a long-absent species. These are only two examples of species that used to widely populate Zena and other areas of the Willamette Valley. Today, however, humans and invasive species threaten their habitat and have reduced their numbers to small percentages of what they used to be. When reflecting on these projects, Professor Bowersox expressed, "I hope I live to see Fender's blue butterfly at Zena." Through these examples Professor Bowersox voices his hope for Zena to serve as a sustainability field station, a place where species are preserved and students can explore the value of ecological restoration. With its central location amidst other farms and small landowners, Professor Bowersox hopes Zena can serve as an example and resource for people to learn about sustainable land management.

Professor Bowersox undeniably realizes the potential for Zena within the Willamette community and understands that developing its potentially many uses will require a long learning process. "As cheesy as it might sound," he states, "[Zena] could be the shining jewel for what Willamette is, in the sense that when students come here this is one of the things that they can really engage in that makes Willamette really distinct." Bowersox believes offering prospective students tours of the property could enhance this image of Zena. When describing Zena, Bowersox says he has two versions. First is the "canned" speech in which he says, "Zena farm and forest is our 305-acre sustainability station where we have a 5-acre farm and 300-acre forest. We have student programs and run programs for the community [with] hands-on learning experiences, and students hopefully feel like they contribute to the space."

The second version exposes the deeper connection Professor Bowersox feels to Zena. From the grey skies that cover fallen leaves and wilted tomato stalks in the winter to the vibrant yellow flowers and bubbling water that rise in the spring, Professor Bowersox feels that Zena is a place of beauty and freedom. "It is a place," he explains, "that sometimes makes me a little emotional, because I believe it is a place that people put down roots in. You go out there and pretty soon you start building something or ripping something out or planting something there, and you sit there and you think 'This is my place.' It's all about sense of place."

Coming Full Circle

Starting with the burning practices of the Kalapuya, to the establishment of organic agriculture, to the commitment of sustainable forestry, current practices at Willamette have in some ways come full circle from the actions of past residents. In many ways we are also evolving and progressing, and the standards set by Willamette's motto and values are what moves us foreword. These values set a precedent for actions by community members, but they are acted out through each individual's interpretation of their meanings. Here we come full circle to the benefit of storytelling. Derived from the experiences and emotions developed in an individual's sense of place, through sharing stories we are able to relieve the possibility of tension and look forward to the future of Zena together. From students and faculty to Salem residents, each story takes the voices of Willamette University's community members and integrates them with voices of people from the surrounding community. As Professor Bowersox indicated, this process is not one that will happen quickly, nor should it. It is a learning opportunity. Through exploring this opportunity, the future of Zena will be one open to people of all walks of life. A place where individuals can come together and experience nature through the soil, the oak trees and maybe even the Fender's blue butterfly, in a way that allows us to put down roots and feel our sense of place.

Notes

1. "Initiatives: Mission & Motto," Willamette University.
2. Joe Bowersox, in discussion with Heather Smith, 2 November 2012.
3. Basso, *Wisdom Sits in Places*, 55; Lang, "From Where We Are Standing, 83.
4. Stephen Thorsett, in discussion with Heather Smith, 6 November 2012.
5. Peter Henry, in discussion with Heather Smith, 24 October 2012.
6. Matthew Nelson, in discussion with Heather Smith, 18 October 2012.
7. Rebecca Dobkins, in discussion with Heather Smith, 17 October 2012.
8. Bowersox, in discussion with Smith.

Epilogue

Bob H. Reinhardt

When asked to reflect on the experience of writing this book, the student-authors noted that the class did much more than provide an opportunity to learn about Zena itself. To many, the project revealed what environmental history is all about—something that became obvious very early in the course, as Philip Colburn recalls: "After taking the class for a few weeks, I realized that it wasn't just about the history of the places; this class was about the history of people connecting with the world around them." The course's particular approach to doing environmental history suggested some of the potential explanatory and analytical power of the field, as Summer Tucker explains:

> I feel that environmental history is an excellent way to examine the interplay of many different players, but what I find special about it is what you get out of writing one. Writing environmental history, you not only have to understand past dialogues between different forces but also push yourself to have your own dialogue with that history. In the end, that whole process seems to give you a different sense of place than you had before. That's what's unique and exciting about environmental history to me. Once you've engaged in it, you just can't look at that snippet of the world the same way again.

Andrew Spittler had a similar experience:

> Helping write this environmental history of Zena and sense of place widened my view of history: what it is and how it can be written and discussed. Environmental history as a discipline continues to intrigue me, as it is a combination of two passions [environmental science and history] I once thought dissimilar, but now see as inseparable.

Many student-authors remarked on the intensive work that this project required. Lauren Vermilion was invigorated by the challenges of directing her own research and writing: "The total freedom of exploration was at my fingertips and I have never enjoyed any writing project more than this one." Vera Warren, too, found benefits in the hard work: "This project really helped me as a writer and a researcher. It is one thing to research a subject, but to produce a good story using historical facts is a challenge." The student-authors also came to appreciate and value the insights offered by their colleagues: "I was amazed at how much information I could gain just by coming to class and intently listening to my peers," explains Erik Sandersen. Many of the student-authors were particularly struck by what they learned through their interviews with participants in this history. Lauren Henken recalls her time listening to Sarah Deumling: "The stories were entertaining, heartfelt, and gave me a greater appreciation for one's connection to the land. The times spent on the back deck of her house overlooking Zena Forest and listening to Sarah were probably my favorite part of the process."

This was a common sentiment from the student-authors: the project revealed not only others' senses of place, but their own, as well:

> Erica Jensen: "This project helped me think more critically about the way I think and feel about different places and how people (including myself) build the connections that they do."

> Lettajoe Gallup: "Studying and understanding the meaning of environmental history changed my perspective of my everyday sense of place. During the process of writing my chapter I noticed myself taking breaks to acknowledge and think about my own sense of place that I have for my home town, Salem, and places I have visited, including China and Hawaii."

> Amanda McClelland: "I have, for the first time in my life, come to know what it means to have a strong sense of place through living at, exploring, and learning about Zena. . . . I think the only way to have a strong sense of place and a real love for a piece of land is to know about everything that came before you and have faith in everything that will come after you, for only then can we treat the land in a responsible and caring manner."

I think it's fair to say that the student-authors did not expect to be moved by the project in these ways. Nor did I. I knew the project would be a challenge (for me and the students), and I hoped we would learn, as we did, about Zena's past and about the practice of environmental history. But I could hardly imagine the way this course and book would so deeply change my own ways of thinking about how people connect to and interact with their environments. Then again, such surprises are one of Zena's defining features, offered to those who stop, look around, and try to make sense of a remarkable place.

Works Cited

Abbott, Carl, Deborah Howe, and Sy Adler. *Planning the Oregon Way: A Twenty-Year Evaluation*. Corvallis: Oregon State University Press, 1994.

Adler, Sy. *Oregon Plans: The Making of an Unquiet Land Use Revolution*. Corvallis: Oregon State University Press, 2012.

Aikens, Melvin C., Thomas J. Connolly, and Dennis L. Jenkins. *Oregon Archaeology*. Corvallis: Oregon State University Press, 2011.

Allen, John Elliot. *The Magnificent Gateway*. Beaverton: Timber Press, 1975.

Alt, David and Donald W. Hyndman. *Northwest Exposures: A Geologic Story of the Northwest*. Missoula: Mountain Press Publishing Company, 1995.

Backlund, Virgil L., Elwin A. Ross, Patrick H. Willey, Thomas L. Spofford, and Dean M. Renner. "Effect of Agricultural Drainage on Water Quality in Humid Portion of Pacific Northwest." *Journal of Irrigation and Drainage Engineering* 121,4 (July/Aug 1995): 289–91.

Basso, Keith. *Wisdom Sits in Places: Landscape and Language among the Western Apache*. Albuquerque: University of New Mexico Press, 1996.

Beckham, Stephen Dow. *The Indians of Western Oregon: This Land Was Theirs*. Coos Bay, OR: Arago Books, 1977.

Berg, Laura. *The First Oregonians*. Portland: Oregon Council for the Humanities, 2007.

Boag, Peter. "The Calapooian Matrix: Landscape and Experience on a Western Frontier." PhD diss., University of Oregon, 1988.

———. *Environment and Experience: Settlement Culture in Nineteenth-Century Oregon*. California: University of California Press, 1992.

Bonneville Power Administration. "Fact Sheet: Zena Conservation Easement Protects Habitat in Willamette Valley." August 2007.

Bowen, William. *The Willamette Valley: Migration and Settlement on the Oregon Frontier*. Seattle: University of Washington Press, 1941.

Boyd, Robert. *The Coming of the Spirit of Pestilence: Introduced Infectious Diseases and Population Decline among Northwest Coast Indians, 1774–1874*. Seattle: University of Washington Press, 1999.

———, ed. *Indians, Fire, and the Land in the Pacific Northwest*. Corvallis: Oregon State University Press, 1999.

Brosnan, Cornelius. *Jason Lee, Prophet of the New Oregon*. New York: W. W. Norton and Co., 1932.

Bunting, Robert. "The Environment and Settler Society in Western Oregon." *Pacific Historical Review* 64,3 (Aug 1995): 413–32.

Burkes, Fikret. *Sacred Ecology*. Hoboken: Routledge Books, 2012.
Collins, Lloyd. *The Cultural Position of the Kalapuya in the Pacific Northwest*. Master's thesis, University of Oregon, 1951.
Copes-Gerbitz, Kelsey. "Defining the Historical Context of Zena Forest, Salem, Oregon." Undergraduate thesis, Willamette University, 2010.
Confederated Tribes of Grand Ronde. "The Kalapuya: A Wealthy Way of Life." *Smoke Signals*. Grand Ronde, 1999.
Denlinger, R. P. and D. R. H. O'Connell. "Simulations of Cataclysmic Outburst Floods from Pleistocene Glacial Lake Missoula." *Geological Society of America* 122: 678–89.
Evenden, Matthew. "Reflections: Environmental History Pedagogy Beyond History and on the Web." *Environmental History* 14, 4 (October 1, 2009): 737–743.
Federal Cooperative Extension Service. "Oregon's First Century of Farming: A Statistical Record of Achievements and Adjustments in Oregon Agriculture 1859–1958." Corvallis: Oregon State College, [1958].
Feld, Steven and Keith H. Basso, ed. *Senses of Place*. Seattle: University of Washington Press, 1996.
Gibson, James. *Farming the Frontier: The Agricultural Opening of the Oregon Country, 1786–1846*. Seattle: University of Washington Press, 1985.
Glenn, Jerry. "Late Quaternary Sedimentation and Geologic History of the North Willamette Valley, Oregon." PhD diss., Oregon State University, 1965.
Gray, William H. *A History of Oregon, 1792–1849*. Portland: Harris & Holman, 1870.
Hays, Samuel. *A History of Environmental Politics since 1945*. Pittsburgh: University of Pittsburgh Press, 2000.
Hobbs, Gertrude. "Thanksgiving Service Spring Valley Church Speech: The McLench Family History," 1982. Spring Valley Church records. Eleanor Miller private collection, Salem, Oregon.
Hudspeth, Elmer B., Richard F. Dudley, and Henry J. Retzer. "Planting and Fertilizing." Pp. 147–52 in *Power to Produce*. Ed. Alfred Stefferud. Washington, DC: US Government Printing Office, 1960.
Inglis, Julian T., ed. *Traditional Ecological Knowledge: Concepts and Cases*. Ottawa: International Program on Traditional Ecological Knowledge International Development Research Centre, 1993.
"Initiatives: Mission & Motto." Willamette University. http://www.willamette.edu/about/initiatives/mission_motto/index.html.
Jacobs, Melville. *Kalapuya Texts*. Seattle: University of Washington Press, 1945.
Jette, Melinda. "Kalapuya Treaty of 1855." *Oregon Encyclopedia*. http://www.oregonencyclopedia.org/entry/view/kalapuya_treaty/.

Juntunen, Judy Rycraft, May D. Dasch, and Ann Bennett Rogers. *The World of the Kalapuya: A Native People of Western Oregon*. Philomath: Benton County Historical Society and Museum, 2005.

Kieley, J. F. *A Brief History of the National Park Service*. US Department of the Interior. Last modified 16 June 2003. http://www.cr.nps.gov/history/online_books/kieley/index.htm.

Knaap, G. J. and C. Nelson. *The Regulated Landscape: Lessons on State Land Use Planning from Oregon*. Cambridge: Lincoln Institute of Land Policy, 1992.

Lang, William. "From Where We Are Standing: The Sense of Place and Environmental History." Pp. 79–94 in *Northwest Lands, Northwest Peoples: Readings in Environmental History*. Ed. Dale Goble and Paul Hirt. Seattle: University of Washington Press, 1999.

Leonard, H. Jeffrey. *Managing Oregon's Growth: The Politics of Development Planning*. Washington, DC: Conservation Foundation, 1983.

Leopold, Luna. *Water, Rivers and Creeks*. Sausalito: University Science Books, 1997.

Lewis, Michael. "'This Class Will Write a Book': An Experiment in Environmental History Pedagogy." *Environmental History* 9,4 (October 1, 2004): 604–619.

Loewenberg, Robert. *Equality on the Oregon Frontier: Jason Lee and the Methodist Mission, 1834–43*. Seattle: University of Washington Press, 1976.

Mackey, Harold. *The Kalapuyans: A Sourcebook on the Indians of the Willamette Valley*. Salem: Mission Mill Museum Association, 1974.

MacPherson, Hector and Norma Paulus. "Senate Bill 100: The Oregon Land Conservation and Development Act." *Willamette Law Journal* 10 (1973–1974): 414–21.

Mazzocchi, Fulvia. "Analyzing Knowledge as Part of a Cultural Framework: The Case of Traditional Ecological Knowledge." *Environments* 36,2 (2008/9): 39–57.

Medler, Jerry and Alvin Mushkatel. "Urban-Rural Class Conflict in Oregon Land Use Planning." *Western Political Quarterly* 32,3 (Sept 1979): 338–49.

Nash, Roderick. *Wilderness and the American Mind*. New Haven: Yale University Press, 1967.

National Atlas of the United States. "What is the PLSS?" Last modified 14 Jan 2013. Accessed April 4, 2012. http://www.nationalatlas.gov/articles/boundaries/a_plss.html.

Northwest Power and Conservation Council. "Bonneville Power Administration, History." Last modified June 2010. http://www.nwcouncil.org/history/BPAHistory.asp.

Northwest Power and Conservation Council. "Council Quarterly, Winter 2008." http://www.nwcouncil.org/library/cq/2008winter.pdf.

Oregon Provisional and Territorial Government. Land Claim Records. Oregon State Archives.

Oregon Water Resources Department. *Willamette Basin Reservoir Study: Criteria and Discussion of Existing and Base Conditions.* 1997.

Orr, Elizabeth, William Orr, and Ewart Bladwin. *Geology of Oregon.* Dubuque: Kendall/Hunt Publishing Company, 1976.

Pacific Northwest Electric Power Planning and Conservation Act. Public Law No. 96-501, S. 885, 1980.

Pickrell, William S. Letter to Sanford Watson. *The Quarterly of the Oregon Historical Society* 16 (March 1915): 61–63.

Polk County. *Comprehensive Plan, 1974.* June 1974.

Powers, W. L. and T. A. H. Teeter. *The Drainage of "White Land" and Other Wet Lands in Oregon.* Corvallis: Oregon Agricultural College Experiment Station, 1916.

Price, Don. *Ground Water in the Eola-Amity Hills Area; Northern Willamette Valley, Oregon.* Washington, DC: US Government Printing Office, 1967.

Price, Don and Nyra Johnson. *Selected Ground Water Data in the Eola-Amity Hills Area, Northern Willamette Valley, Oregon.* Washington, DC: US Government Printing Office, 1965.

"Purvine-Walker Papers." Eleanor Miller private collection, Salem, Oregon.

Rinkevich, Sarah, Kim Greenwood, and Crystal Leonitti. "Traditional Ecological Knowledge for Application by Service Scientists." US Fish and Wildlife Service. Last modified February 2011. http://www.fws.gov/nativeamerican/graphics/TEK_Fact_Sheet.pdf.

Robbins, William. *Landscapes of Conflict: The Oregon Story, 1940–2000.* Seattle: University of Washington Press, 2004.

———. *Landscapes of Promise: The Oregon Story, 1800–1940.* Seattle: University of Washington Press, 1997.

Rothman, Hal. *The Greening of a Nation? Environmentalism in the US since 1945.* Orlando: Harcourt Brace, 1998.

Salem Pioneer Cemetery. "Sanford Watson." Cemetery record. http://www.salempioneercemetery.org/records/pf_display_record.php?id=7070.

Salix Associates. "The Zena Property: Rare Plant and Butterfly Survey and Blackberry Mapping." October 2008.

Savoca, Maria. "Land Use and Vegetation Change at Zena Forest, Oregon, 1935–2005." Undergraduate thesis, Willamette University, 2009.

Sims, Berry. *Forest Stewardship Plan for the Willamette University Forest at Zena.* Portland: Trout Mountain Forestry, 2008.

Stegner, Wallace. *Where the Bluebird Sings to the Lemonade Springs: Living and Writing in the West.* New York: Random House, 1992.

Stevens, Michelle. "Common Camas." US Department of Agriculture. Last modified 31 May 2006. http://plants.usda.gov/plantguide/pdf/cs_caqub2.pdf.

Stoll, Mark. *Protestantism, Capitalism, and Nature in America*. New Mexico: University of New Mexico Press, 1997.

"Summer Institute in Sustainable Agriculture." Center for Sustainable Communities. http://www.willamette.edu/centers/csc/zena/.

Taylor, George. "Climate of Multnomah County." Oregon Climate Service. http://www.ocs.orst.edu/county_climate/Multnomah_files/Multnomah.html.

Trust for Public Land. "Mission & History." http://www.tpl.org/about/mission/.

———. "Zena Forest Property." http://nnrg.org/files/Zena_Brochure.pdf.

"The University Forest at Zena: Mission and Goals." http://www.willamette.edu/centers/csc/zena/forest/mission/index.html.

Tuan, Yi-fu. *Space and Place: The Perspective of Experience*. Minneapolis: University of Minnesota Press, 1977.

US Department of Agriculture Natural Resources Conservation Service (USDANRCS). "Web Soil Survey." Accessed December 13, 2012. http://websoilsurvey.nrcs.usda.gov/app/WebSoilSurvey.aspx.

US Soil Conservation Service. *Willamette Valley Drainage Guide*. Portland, OR: Soil Conservation Service: 1977.

Usher, Peter J. "Traditional Ecological Knowledge in Environmental Assessment and Management." *Arctic* 53,2 (June 2000): 183–93.

Walker, Peter and Patrick Hurley. *Planning Paradise: Politics and Visioning of Land Use in Oregon*. Tuscon: University of Arizona, 2011.

Walsh, Megan K., Cathy Whitlock, and Patrick J. Bartlein. "1200 Years of Fire and Vegetation History in the Willamette Valley, Oregon and Washington, Reconstructed Using High-Resolution Macroscopic Charcoal and Pollen Analysis." *Paleogeography, Paleoclimateology, Palaeoecology* (August 2010): 273–89.

White, Richard. *"It's Your Misfortune and None of My Own": A New History of the American West*. Norman: University of Oklahoma Press, 1991.

———. *Land Use, Environment, and Social Change*. Seattle: University of Washington Press, 2000.

Willamette River Basin Memorandum of Agreement Regarding Wildlife Habitat Protection and Enhancement between State of Oregon and the Bonneville Power Administration. October 22, 2010. Attachment B. Project Selection Criteria.

Willamette River Basin Memorandum of Agreement Regarding Wildlife Habitat Protection and Enhancement between State of Oregon and the Bonneville Power Administration. October 22, 2010. Attachment C. List of Completed Wildlife Projects.

Willamette River Basin Memorandum of Agreement Regarding Wildlife Habitat Protection and Enhancement between State of Oregon and the Bonneville Power Administration. October 22, 2010. Attachment G. Operations and Maintenance ODFW Baselines 5.

Women's Missionary Society. "An History of the Spring Valley Church at Zena, Oregon," 1958. Spring Valley Church records. Eleanor Miller private collection, Salem, Oregon.

Wright, Albert. "An Economic and Biographical History of Heirloom Apple Trees at Zena Farm." Undergraduate thesis, Willamette University, 2010.

www.ingramcontent.com/pod-product-compliance
Lightning Source LLC
Chambersburg PA
CBHW051103160426
43193CB00010B/1300